MAKING
Knowledge
COMMON

Colin Lankshear, Michele Knobel,
Chris Bigum, and Michael Peters

General Editors

Vol. 16

PETER LANG
New York • Washington, D.C./Baltimore • Bern
Frankfurt am Main • Berlin • Brussels • Vienna • Oxford

Lesley Farrell

MAKING
Knowledge
COMMON

LITERACY & KNOWLEDGE AT WORK

PETER LANG
New York • Washington, D.C./Baltimore • Bern
Frankfurt am Main • Berlin • Brussels • Vienna • Oxford

Library of Congress Cataloging-in-Publication Data
Farrell, Lesley.
Making knowledge common: literacy and knowledge at work / Lesley Farrell.
p. cm. — (New literacies and digital epistemologies; v. 16)
Includes bibliographical references and index.
1. Communication in organizations. 2. Knowledge management.
3. Employees—Training of. 4. Workplace literacy.
5. Communication of technical information.
I. Title. II. Series.
HD30.3.F38   658.4'038—dc22   2006000065
ISBN 0-8204-6761-8
ISSN 1523-9543

Bibliographic information published by **Die Deutsche Bibliothek**.
Die Deutsche Bibliothek lists this publication in the "Deutsche
Nationalbibliografie"; detailed bibliographic data is available
on the Internet at http://dnb.ddb.de/.

Cover design by Sophie Boorsch Appel

© 2006 Peter Lang Publishing, Inc., New York
29 Broadway, New York, NY 10006
www.peterlang.com

All rights reserved.
Reprint or reproduction, even partially, in all forms such as microfilm,
xerography, microfiche, microcard, and offset strictly prohibited.

Contents

Preface: That's enough about them, let's talk about me vii

Acknowledgments xi

Chapter One: Making knowledge common 1

Chapter Two: Workplaces/workspaces 25

Chapter Three: What counts as knowledge in
   knowledge economies? 45

Chapter Four: Solving problems by the book 67

Chapter Five: Solving problems in cyber-place 91

Chapter Six: Learning how to be: the textual production
   of working identities 113

Chapter Seven: Workplace educators 'working' knowledge
   in the Knowledge Economy 133

Afterword: If workplace literacy education is the answer,
   what is the question? 153

References 157

 # Preface

That's enough about them, let's talk about me . . .

> What did the postmodern interviewer say to the interviewee?
> That's enough about you, let's talk about me.
>
> (McLeod)

The first time I walked into the Exmouth Plant I felt the noise reverberate through my feet and legs, finally squeezing the breath from my chest. Speaking wasn't possible, but, then, hearing wasn't possible either. I'd worked in a printing factory years before, and I'd done night shift on a production line, and they were noisy enough, but I'd never had my body taken over by noise like this. People knew who I must be—the researcher from the university who was going to be around for a while, if she could stand it—so they mimed 'hello' and went on with their work. Bill gave me earplugs to wear, which dulled the noise but not the vibration. He was expecting a male researcher, I think, but, if he was, he got over his disappointment and took me around to introduce me to every huge machine. He told me what could go wrong, and what you needed to know and be able to do to fix it, or, at least, to get it going long enough to complete an overdue order. I, of course, couldn't hear a word, but Bill was 'wired' with microphone and tape recorder. I listened to the tape in the lunchroom and tried to match the commentary with the appropriate machine so he wouldn't have to tell me again.

I was there to investigate the concept of 'competence' from a language point of view[1]. Bill did not have a high opinion of the nationally mandated competence frameworks I was interested in, and there didn't seem to be a lot of language going on, so I wasn't sure, on that first day, how long my stay would be. As it turned out, I was there for over eight months initially, and I returned from time to time over the next two years. With Miriam, my research assistant, I went to all the Action Learning Team meetings, and to various formal and informal problem solving meetings, and taped them all. I interviewed all the members of the Action Learning Team (some several times) and various members of senior management. I hung around the Warping Shed, the Weaving Shed and the Mending Room. I sat at a mending table at the Harbor Mill while Mary placed the burler, the scissors and the sewing needle in my hand and taught me how to mend, and I crawled under a warping machine while Bill explained how humidity affected tension, and what to do about it. I learned about the problems of sourcing yarn, the difficulty of getting consistency in fabric, and what happened when the big automotive companies changed their minds at the last minute. I heard what it feels like to work where your mother and aunts worked but to rise to a position that those women would never have aspired to. I heard what it feels like to feel you are being pushed out.

Over time, as I watched the people at the Exmouth Plant and the Harbor Mill grapple with increasingly insistent demands to turn what they knew and what they did into words, I became less and less satisfied with the idea of 'competence' as a way of thinking about knowledge at work. At the same time, I became more and more sure that the term 'workplace literacy' failed to signal the social transformations associated with changing reading and writing and talking at work in Australia, where I was, and in the world that we were part of. From Sally and Margaret, workplace educators, I heard what it was like to try and encourage, persuade, cajole, and even try to force people into coming to meetings, talk about work, and write things down, even when the viability of the company, and of their jobs, depended on it.

Sally and Margaret's troubled reflections led me to the Workplace Educators Project[2]. They had talked about why people didn't want to, or couldn't, write at work, when they were clearly literate, and about how relationships with their students, for the first time in their professional lives, were uncomfortable, sometimes openly hostile. I wondered if other workplace eductors were feeling the strain. As it turned out, they were. The workplace educators I talked with in the Workplace Educators Project were relieved to talk, and to hear 'it isn't just me'. Regardless of their backgrounds (corporations, trade unions, vocational colleges, consultancy) they were caught up in a phenomenon that was redefining their roles, their relationships and their understanding of what on earth they were supposed to be doing in workplaces. They relished the opportunity to talk and to try and make sense of what they were experiencing. For me their stories were compelling accounts of what it is like to try to make knowledge networks connect when you didn't know who you were connecting with or why you were connecting. They were all teaching people to read and write and talk differently, but none of them were called literacy teachers and most of them had no language or literacy training. Their students seemed, at best, ambivalent about learning what they had to teach them and sometimes overtly angry with them. They all expressed an uncomfortable feeling that they were 'on the wrong side', although they didn't know how they got there, or where 'the right side' was. They were isolated in workplaces and moving around all the time. They didn't have time to look up and see that other workplace educators were feeling the same, and it was hard for them to get a sense of any particular organization and the internal and external forces that were shaping it. For most of them, globalization was something that happened 'out there', they did not see themselves as 'doing globalization'. But that was how I came to see their work. As they struggled to get people to fill in the pro forma and keep minutes of meetings, to solve problems using fishbone diagrams and to read and annotate 'The Manual', they were, as far as I could see, trying to create the textual ligatures that joined up the global Knowledge Economy.

So this book came to be about the textual practice of knowledge at work. It is a book that could not have been written without the stories of the people from Exmouth Plant, the Harbor Mill and the Workplace Eductors Project, but this is my story, not theirs. I have chosen from hundreds of hours of transcripts the small sections to be used here, I have chosen how to analyze them and I have interpreted them. I have drawn pictures of people that may bear little relation to the way they see themselves or the ways their workmates and family see them. I have described their workplaces, and I have drawn conclusions about people and relationships. I don't think there is too much to be gained by being precious about this; this book says more about me than it does about them. Someone else would have seen it differently. Indeed, I would have had seen it differently if I'd done it earlier, or later. All the same, I hope I have made it clear that I like and respect, and am enormously grateful to, the people who welcomed me into their working lives. I hope, if they ever read this book, they will recognize something of themselves and their experiences in it, and think it was worth it.

NOTES

1. The Textual Practice of Competence in the Textile Clothing and Footwear Industry Project was funded by a grant from the Australian Research Council. Names of people, places and organizations have been changed to protect anonymity in all the projects quoted.
2. The Workplace Educators Working Knowledge in the Global Economy was funded by a grant from the Faculty of Education, Monash University.

Acknowledgments

The making of a book begins so long before the words start to bleed onto the page that it is impossible to acknowledge, or even to know, everyone who has had a part in its production. Here I will mention only a very few of the people who have contributed, in various important ways, to this one: Catherine Beavis, Miriam Faine, Michael Farrell, Peter Freebody, Virginia Freebody, James Gee, Bernard Holkner, Paul James, Lesley Johnson, Barbara Kamler, Colin Lankshear, Allan Luke, Heather Phillips, Terry Threagold, Terri Seddon and Andy Spaull.

I have been thinking about these ideas for the last few years. During that time some of the material in this book has appeared in a preliminary form in other places, specifically:

'Workplace education and corporate control in global webs of interaction.' Pp. 479–493. 2004. vol. 17:4. *Education and Work* http://www.tandf.co.uk; 'Negotiating knowledge in the knowledge economy: workplace educators and the politics of codification.' Pp. 201–214. 2001. vol. 23:2. *Studies in Continuing Education* http://www.tandf.co.uk; 'Ways of doing, Ways of being: Language, education and working identities.' Pp.18–36. 2000. vol. 14:1. *Language and Education*. 'The "New Word Order": workplace education and the textual practice of globalization' Pp. 57–74. 2001. vol. 9:1. *Pedagogy, Culture and Society*.

CHAPTER ONE

Making knowledge common

> Common knowledge is a phenomenon which underwrites much of social life. In order to communicate or otherwise coordinate their behavior successfully, individuals typically require mutual or common understandings or background knowledge. Indeed, if a particular interaction results in 'failure', the usual explanation for this is that the agents involved did not have the common knowledge that would have resulted in success.
>
> (VANDERSCHRAAF)

INTRODUCTION

Knowledge is a problem in the global economy. It is true that, for all companies, large, small and medium-sized, the production of knowledge is core business. The free communication of knowledge internally and across global networks of interaction is absolutely fundamental to the health of any company, and the commodification of knowledge is critical in any organization's economic viability. But, the work practices that nurture the production and communication of knowledge are not the same work practices that establish and maintain control of the knowledge, nor are they the work practices that allow the exploitation of knowledge as an economic product traded on a global market. In fact, the social practices that cultivate knowledge seem to be those that work against corporate control; they promote and rely on the independent judgment of individuals and groups of workers. Companies are faced with the dilemma of organizing themselves in such a way that they 'capture knowledge without killing it' (Duguid 2000).

Not so long ago, this seemed to be a problem confined to the management of elite 'knowledge workers', that select group of highly creative people who produce the innovative ideas and designs that power the global economy. Now the arbitrary distinction between 'knowledge workers' and the rest is a much more difficult distinction to sustain. A feature of contemporary companies is the primary role that written and electronic texts (and the spoken texts they generate) play in joining up people and practices in globally dispersed organizations, and the global networks of interaction of which they are a part. Now most workers, even in the smallest and most local of companies, are involved in some way in the innovative and creative processes of textualizing knowledge at work. Knowledge is not transmitted *by* texts, it is made and transformed *in* and *with* texts, and the people who make and use the texts are real knowledge workers, in the sense that they produce the knowledge that makes the Knowledge Economy happen.

Workplace educators are faced with the challenge of helping people and organizations coordinate and negotiate the work practices and working relationships, realized in textual practices, which create working knowledge. Part of the challenge is, of course, helping people create the new kinds of texts that global networks of interaction demand. In many respects this is an exciting challenge, just the kind of challenge that workplace educators take delight in meeting. Another part of the challenge is more difficult for workplace educators to identify, understand, and negotiate. The new textual practices of work are not innocent. While they make and transform knowledge they also regulate and control knowledge, and the people and work practices that produce it.

Workplace educators are recruited, not merely, or even primarily, to the project of developing the literacy practices of workers; they are recruited to the project of developing the texts that manage knowledge and people in globally distributed companies. They help create the textual web that tames knowledge, and, perhaps, the people who produce it; they help create the textual web that 'captures knowledge without killing it'. People, very often, resist being tamed; they defy, subvert or co-opt textual practices to their own purposes. For workplace educators, navigating workplace education in the Knowledge Economy can be a tricky business.

In one way or another, making knowledge and making it common is the challenge of the Knowledge Economy. This book is an exploration of how people use language to make the global Knowledge Economy happen. It started with my attempt to understand how a group of people who work for a textile company in Victoria, Australia, come to adopt (or adapt, or resist) new ways of reading, writing, listening and speaking at work. Initially it was a fairly straight forward analysis of the 'textual practice of competence'—an attempt to identify the literacy practices textile workers needed to demonstrate if they were to be deemed 'competent' in the new industrial environment of industry, national and international standards. As I became more familiar with the people who worked in the textile manufacturing plants I became aware of how little I understood about the work people did, who they did it with, the reasons they did it that way, and why those ways were changing. As I learned more about the factory, more about the textile and automotive industries, and more about the global supply chains that produce the global product, I became less interested in whether people could meet the literacy demands of the workplace and more interested in why those demands were being made in the first place. Even to me (an interloper who researched language and social change and talked and wrote for a living), and certainly to them, there seemed to be an awful lot of reading, writing and talking going on. The presence of workplace educators to coach and scaffold suggested that this was new and important textual practice, and that the textile workers could not learn to do it by themselves. My focus shifted from how literate the textile workers were to the texts themselves: what people were reading, what they were writing, and what kinds of things they were being asked to talk about and to listen to. The texts they were making (or approximating, adapting, or refusing to make) were texts that linked them and their organization to a global network of knowledge production and knowledge regulation.

And so this book has come to be about how they, and we, as individual workers and groups of people in factories and offices and workshops, use textual practice to make knowledge common in the global economic arena, and what is at stake when we do.

For the most part this book is concerned with specific people going about their everyday work, but in this chapter I want to sketch a backdrop to that practice, to give an idea of the global conditions which their work constitutes and which constitutes their work. I start with a discussion of three ways of thinking about 'common knowledge' in the global workplace: knowledge held in common, knowledge made in communities, and commonplace common knowledge.

COMMON KNOWLEDGE MATTERS

Knowledge in common

Common knowledge has always mattered at work. The Craft Guilds of medieval Europe built bodies of knowledge that were developed and used 'in common'. Local alliances of master craftsmen developed, formulated, and regulated, how to build a dry stone wall, or make a shoe, or bake a loaf of bread. Apprentice masters took apprentices into their workshops so that the apprentices could watch and learn beside them. The fundamental purpose of the apprenticeship system was to induct teenage boys into the discourses and practices that constituted the common knowledge of their trade. Craft Guilds were established to build knowledge, to preserve it, and to pass it on to the next generation. At the same time, the Guilds protected their craft knowledge by applying rigorous standards to production processes and products, thereby ensuring effective control over technological innovation and work processes (Epstein 1998).

The Craft Guilds weren't called 'mysteries' for nothing. Part of the indenture contract of an apprentice demanded that the apprentice work in his master's workshop for five to seven years, do as his master told him, and keep his master's occupational knowledge to himself. The Craft Guilds were about making and building knowledge but, like companies today, they were also about regulating and controlling it. While the Guilds ensured the quality of products by applying commonly agreed standards, they also ensured that the knowledge of the trade remained secret, and didn't extend beyond the members of the local Guild, and that the numbers entering the occupation were regulated to ensure that prices (and consequently incomes) were maintained (Berezin 2003). Craftsmen needed knowledge in common to get things done and to solve common problems, but they wanted to protect their knowledge as an asset too. The challenge of the mediaeval Craft Guilds was to build and promote occupational knowledge and innovation, and ensure a supply of skilled craftsmen, while ensuring that the occupational knowledge base was regulated and access to it controlled.

Common occupational knowledge was 'local knowledge'. It was the product of local activity and effort and was viewed as property held in common by the local community. Geographical boundaries mattered, and the penalties for breaching them were harsh and strictly policed. In medieval Florence, for instance, any craftsman who, possessing the secrets of particular manufacturing process, fled the town for any reason, had to be tracked down and killed (Betcher 2004). Common knowledge was, simultaneously, the relationship which bound the members of the Guild to each other and the commodity which could irrevocably dissolve that bond.

Craft Guilds were outlawed in most of Europe by the late eighteenth century; their monopolistic practices were seen as a hindrance to the development of the new Industrial Economy. The new Knowledge Economy, however, wants them back. The Craft Guilds of medieval Europe have been self-consciously resurrected by information technology specialists. They have established the Systems Administrators Guild, the Silicon Valley Web Guild, the Html Writers' Guild and the Graphic Artists Guild (Benner 2003). The President of the HTML Writers' Guild indicates that the term 'Guild' was chosen deliberately:

> ... in order to look back at the older, medieval-type Guilds. What we liked from that model was the notion of sharing knowledge–that building web design was something of a craft–not purely artistic or purely technical (Benner 2003:186).

Workers in Information Technology industries are not keen to replicate the highly regulatory, possibly monopolistic, character of medieval Guilds. They do, however, often find themselves isolated; they often work alone as contractors and are associated with no particular physical workplace and with no stable group of colleagues. It is difficult to keep up with new technological developments and with other developments in their industries. These new Craft Guilds are concerned with 'finding a world in common' (Smith 1999) in which they can share occupational and industry knowledge and solve problems in a context in which a certain amount of shared occupational knowledge, practice and interest can be taken for granted. Benner, in his study of the Guilds of Silicon Valley, argues that they can best be understood as learning communities:

> Perhaps most importantly they provide an important learning infrastructure, helping their members increase their own skills and learning opportunities over time. In the environment of rapid change and volatility that characterizes the information-based economy of Silicon Valley, it is this ability to help their members deal with rapid change that is most critical (2003:191).

Common occupational knowledge is knowledge that is made and held 'in common'; it identifies knowledge as a commonly held asset base and it supports the development of a knowledgeable workforce in conditions which militate against it.

Common knowledge and common practice

Common knowledge is also associated with community, and with the common views, values and practices that bind communities together. In the past, people's routine, informal connections with each other as they engaged in common practice could be taken for granted in successful businesses; the importance of a community in generating and maintaining knowledge was so obvious it didn't need to be stated. This is no longer the case. In the globally distributed workplace the nurturing of communities which share work practices is a deliberate, and heavily promoted and documented, business strategy (Senge 1991; Nonaka and Takeuchi 1995; Lindstaedt 1996; Gee 1997; Ezzamel and Willmott 1998; Engestrom 1999; Wenger 1999; Brown and Dugaid 2000; Duguid 2000; Gee 2000; Hildreth and Kimble 2000; Lesser, Fontaine et al 2000; Snyder 2000; Hildreth and Kimble 2004).

The driver in the development of the new high-tech Craft Guilds is the need to create 'a community' in which people can make, learn and protect new knowledge. This is the driver that is exploited in the familiar idea of 'communities of practice':

> Communities of practice are groups of people who share a concern or a passion for something they do and learn how to do it better as they interact regularly (Wenger 2005 on-line).

Workplace communities are important because knowledge production requires a shared culture; shared stories which underpin shared understandings, values and practices. Collective memory, and stories constructed together over time, are important not just because they get the work done, but because they generate new knowledge (Orr 1990). Workplace communities are important because knowledge production relies on the relationships of trust and commitment to a joint enterprise that the idea of community implies (Prusak and Cohen 2001). A shared culture means that people know one another's stories (sometimes too well) and use them as resources, and they can risk improvisation, confident in the knowledge that their colleagues will support them, and bail them out if they need to. Many fast capitalist companies embrace the idea of 'communities of practice' as a way of simultaneously making and taming knowledge (Duguid 2000). Like the medieval workshops, they are under pressure to make and use knowledge but also to keep it safe from marauding competitor organizations. The problem for organizations, of course, is that communities of practice are good for producing knowledge, and for protecting it as a common asset, but they are not amenable to control.

Commonplace common knowledge

The success of communities of practice in workplaces relies, however, on common knowledge that is both broader and more intimate than occupational knowledge. The term 'common knowledge' is commonly used, and relevant to any discussion of global companies. The *Stanford Encyclopedia of Philosophy* defines common knowledge as:

> a phenomenon which underwrites much of social life. In order to communicate or otherwise coordinate their behavior successfully, individuals typically require mutual or common understandings or background knowledge. Indeed, if a particular interaction results in "failure", the usual explanation for this is that the agents involved did not have the common knowledge that would have resulted in success (Vanderschraaf Summer 2002 on-line).

This understanding of common knowledge focuses on the kinds of background, personal, cultural, political and social knowledge that is frequently taken for granted, and, therefore, not stated explicitly, in the hurly burly of the contemporary workplace.

Since it is people like you and me who make common knowledge, it shouldn't be surprising that there may be little agreement about what counts as common knowledge in any situation. This may be because our contexts are quite different. When we speak on the phone, I (in Melbourne) might 'know' that night is about to fall, you (in New York) might 'know' that the sun has just risen. When we read a measurement of 56, I might 'know' it means inches, you might 'know' it means centimeters. Or it may be because different interests are served when a particular set of assumptions is considered common knowledge, and therefore not up for discussion. I might 'know' that dentists should earn more than kindergarten teachers; you might 'know' that sports people should earn more than doctors. Having our context accepted as the default context, and our beliefs accepted as 'common knowledge' and not easily contestable, puts one of us in a powerful position.

All these kinds of 'common knowledge' are challenged when people work in a global Knowledge Economy. It is easy to talk about the global Knowledge Economy as if it is a single monolithic entity that exists independently of the work we do and the people we do it with. In this book I'm working with a different idea of a global Knowledge Economy, one that focuses on the micro-processes of globalization and the social practices that constitute them. In the section that follows I'd like to explain how I see knowledge in the Knowledge Economy and what I mean when I use the terms 'Knowledge Economy' and 'knowledge worker'.

WORKING IN A GLOBAL KNOWLEDGE ECONOMY

The common knowledge that counts

People make global knowledge economies happen, and they do so in an uncertain, unpredictable and frequently uncharted economic world. For the most part, we don't have the luxury of waiting till a homogenous, technologically invincible and comprehensively regulated global economy is up and running. We function in our own particular corners of the Knowledge Economy on a case-by-case basis, using the people, data, information, documentation, knowledge, infrastructure and technical resources we have available. When the territory is unfamiliar, the technology breaks down, or acts of god or man interrupt the theoretically seamless communication networks we rely on, we improvise, mobilizing our professional and personal networks to make the knowledge (about processes, systems, people and places) we need to finish the job. This kind of workplace knowledge-making sits alongside more explicit, formal vocational and professional knowledge; it is not new nor is it peculiar to work in globally networked economies. Improvisation ('filling in the gaps') is one of the most important and extensively documented forms of organizational knowledge-building in the literature, and one of the hardest to pin down and teach (Granovetter 1973).

What distinguishes contemporary global economies from those that have gone before is that all kinds of knowledge production (formal knowledge-making and improvisation) happen across geographically and temporally dispersed sites. Neither global corporations nor local companies can rely exclusively, or even largely, on the embodied knowledge and shared local work practices that supported medieval craft workshops and local Guilds, and have supported production for centuries. Instead they must rely on people being able to initiate connections and sustain relationships across and within workplaces, cultures, and organizations. To the extent that people all over the world can generate the right conversations, compose the persuasive Emails, make the timely phone calls, complete the critical paperwork, solve the problems and create the makeshift knowledge and practices that span the process and communication chasms, products and services do get to the right destination in the global market 'just in time', and the global economy continues to happen.

Of course we know that globalization exerts conflicting pressures. The effects of accelerating time and diminishing space (the almost instantaneous movement of data around the globe, for instance) are well documented identifying characteristics of the global economy (Lash and Urry 1994; Soja 2000). They exert pressures towards the standardization and homogenization of products, of services, of cultures, and of the ways people do their work. However, at the same time as they exert pressures for standardization and conformity, the acceleration of time and the contraction of space puts new pressures on individuals and groups of workers to act independently and creatively, to make new rules and break old ones. Decisions have to be made more quickly than ever, and there are many situations where individuals and groups of workers are expected to use their own knowledge and judgment in their local settings to develop innovative solutions to immediate problems without reference to Head Office authority. This can be bewildering to those of us operating in these contexts, precipitating existential crises over who we are supposed to 'be' at work—the compliant, meticulous follower of regulations or the creative, autonomous free spirit. At work, it seems, people need to be both, simultaneously compliant and creative. While globally distributed supply chains seem to demand unquestioning adherence to globally enforceable regulatory frameworks, global companies are increasingly reliant on the knowledge-building capacities of the people who work in them for this kind of creativity.

Standardization and improvisation

Global knowledge economies need standard knowledge and they need niche knowledge, and both kinds of knowledge need common knowledge. On the one hand, the challenge is to identify, transmit and transfer 'standard' knowledge; to come to an agreement across plants, offices, boardrooms and companies about what will constitute 'common knowledge' about products and processes, how that knowledge will be produced and about the means of ensuring that common knowledge is commonly understood in disparate contexts. This is a critical requirement of any shared enterprise, and difficult enough to establish at the best of times, when people are in routine contact in familiar shared physical contexts. The need to standardize products for global markets (to make sure that a luxury car assembled in South Africa is identical to the same make and model assembled in Germany for instance) makes it critical in contemporary workplaces. The rapid development of new knowledge (innovation) makes the processes of sharing and legitimating knowledge across geographical and corporate boundaries urgent. It has led to various attempts by global companies to exhaustively document 'best practice' and make exactly the same technical and process knowledge available, at exactly the same time, and to demand that the same knowledge-producing practices be utilized, across all the companies involved in the production of an item. Various more or less comprehensive regulatory frameworks (Quality Standards Frameworks, Standard Operating Procedures, Business Process Manuals etc) are designed to document 'common knowledge', and to ensure that identical knowledge and practice is shared amongst all the people involved in the production of an item, regardless of where they work. Absolute compliance with the most important of these regulations is critical if a company is to sustain its place in a global supply chain. These frameworks may mandate organizational practices (like the adoption of cross-functional teams) that challenge established local ways of working at every level, may diminish the value of many locally produced goods and services, and may seem to create a wholesale reduction in the autonomy and authority people can exercise when they work.

On the other hand, the challenge of 'making knowledge common' is the challenge of improvisation; a challenge not to standardize but to customize. Improvisation happens when small groups of people identify a problem (or opportunity) and together forge the knowledge to meet the immediate need. No matter how hard they try to be, regulatory frameworks, quality manuals and standard operating procedures are not comprehensive; they cannot take full account of local exigencies. What's more, the global economy will not stand still to be regulated, and workers can find that they are faced with a plethora of regulations and corporate and industry sector standards frameworks sitting along side local rules, regulations and standard operating procedures of varying relevance and uncertain currency and legitimacy. At the same time there are significant areas of work which are completely unregulated, where conflicting regulations apply, or where new conditions have outpaced existing regulations so comprehensively that they cannot be made to fit, and they have no guidance on what they should do or how they should do it. Within this context improvisation is at least as important a dimension of knowledge-building as it ever was.

Making knowledge common involves more than following global manuals, and it involves more than the creation of new knowledge to meet specific immediate needs. It involves more than working with geographically proximate colleagues to solve problems. The global economy demands that knowledge is created and used collectively—'in common'. This is not a nebulous ideal, it is an everyday necessity. For most people at work, the most notable feature of the new economy is the relentless demand that they make and maintain productive connections between people, departments and organizations. Castells explains why this demand is so unremitting:

> At the heart of the connectivity of the global economy and of the flexibility of informational production there is a new form of economic organization, the network

enterprise. This is not a network of enterprises. It is a network made from either firms or segments of firms and/or from internal segmentation of firms. Large corporations are internally decentralized as networks.' Small and medium businesses are connected in networks. These networks connect amongst themselves on specific business projects and switch to another network as soon as the project is finished. Major corporations work in a strategy of changing alliances and partnerships, specific to a given product, process, time and space. Furthermore, these cooperations are based increasingly on sharing information (Castells 1996: 10–11).

When Castells says that 'these networks connect amongst themselves on specific business projects and switch to another network as soon as the project is finished' it sounds as if the whole process occurs without human intervention. Of course, this is not the case. The firm is little more than an idea until it is animated by people, and it is the people employed in the firm who establish and sustain the connections with other people that create functioning networks. The capacity of a workforce to establish and sustain productive, knowledge-sharing, connections with a wide variety of other people and organizations is now critical to the success of an enterprise, no matter what its size or business. The capacity to switch between networks, to extract themselves from one and reinsert themselves in another, must, equally become second nature.

In the contemporary workplace people work in formal and informal groups that are geographically and temporally distributed, even within the one company. However, as Castells points out, most products are not produced by a single company; critical components are outsourced and assembled all over the world and production is coordinated through a complex and sophisticated communication system. The least sensitive, and most extensively documented, part of this process is the transmission of data and information; the most sensitive, and least explored, is the negotiation and collaboration that leads to 'new knowledge'—innovation and improvisation that are shared understandings which can be animated at different sites by different people with a common problem and with more or less predictable results. 'Common knowledge' is made by people communicating—more often than not through digitized print and ICT-enabled communication channels—E-mail, telephone, print and virtual workspaces.

Knowledge workers and the Knowledge Economy

Our understanding of what constitutes a 'Knowledge Economy' and a 'knowledge worker' has expanded significantly since Reich's (Reich 1991) identification of knowledge workers as 'symbolic analysts'—an elite group of people who worked in a more or less discrete 'Knowledge Economy', primarily with

Information Communication Technology (ICT)-enabled semiotic systems: data, words and oral and visual representations of ideas in various technologically mediated environments. Reich identifies symbolic analytic work as being that work which involves identifying and solving new problems. It has to do with analyzing, manipulating and communicating through numbers, shapes, words, ideas. This kind of work usually requires a college degree (Reich 2003).

For Reich, symbolic analytic work is done by symbolic analysts. These workers:

> do R&D, design and engineering. Or they're responsible for high level sales, marketing and advertising. They're lawyers, bankers, financiers, journalists, doctors and management consultants (Reich 2003 on-line).

The predicted expansion of this Knowledge Economy, and the predicted decline of the 'old economy', rested on the assumption that the most powerful impact of the new technologies would be as 'products', the development of new commodities to be bought and sold on global markets, and that the most prosperous national economies would be those that focused on 'knowledge work', especially in high-tech areas like biotechnology and nanotechnology.

Not so. Paradoxically, this predication seems to have under-rather than over-estimated the pervasiveness and importance of symbolic-analytic work, and focused attention on only one dimension of the Knowledge Economy. While undoubtedly significant, the most powerful impact of ICTs has not been to introduce a new series of technological products to global markets, no matter how sophisticated and powerful they may be. The most powerful economic impact of ICTs appears to be the way they have transformed the 'old' economy—creating global markets, global regulatory frameworks and global production systems and therefore creating new relationships and alliances, new niches and new ways of understanding and experiencing production, and new kinds of knowledge work. Reich, for example, draws a fundamental distinction between high-income 'knowledge work' and low income 'service work'. He argues that service work cannot easily be mechanized and that it cannot easily be done by workers in other nations; service work he argues must be done in person. In this category he includes cabbies, retail workers, security guards and hospital attendants. While I would be unwise to speculate about the future work practices of cabbies or security guards it is not necessary for me to speculate about retail workers. Already, 'customer service' for large Australian retail organizations is outsourced to nondomestic workers. Call-center operators located in India sympathetically address the problems of customers of a large Australian retail chain. They are trained to conduct their telephone conversations in a culturally sensitive way and to take an active interest in Australian popular culture in order establish productive relationships in which problems can be solved. This work is no longer categorized as 'retail work'; it is part of the operation of a multinational finance company (Pittard, Kowalski et al 2004).

The Knowledge Economy is no longer the domain only of an elite; in one sense or another, everyone inhabits it. This is especially evident in the way that established categorizations like 'industry sector' (finance, retail, service etc) are conflating, acknowledged as inadequate to the task of capturing what businesses do:

> increasingly, knowledge and related intangibles not only make businesses go but are part or all of the 'products' firms offer. Old distinctions between manufactured objects, services and ideas are breaking down (Davenport and Prusak 1998: 47).

Like other categories, the category of 'knowledge work' isn't a very helpful one in this context, either. What is becoming increasingly apparent is that the transformation of the global economy is reframing virtually *all* work as 'knowledge work' in the sense that the active production and application of knowledge keeps *all* businesses operating in ICT-enabled global networks of production. One aspect of this is the increased necessity of reframing existing local knowledge to make it recognizable as 'knowledge', and usable as 'knowledge' at remote sites. This may become the (often unacknowledged) responsibility of the individual worker (the bookkeeper, the weaver, the engineer, the personal assistant). It is more helpful to concentrate on symbolic-analytic work than the symbolic-analytic worker.

While the focus on ICT-enabled workplaces has tended to emphasize the concrete—the transfer of data and information—it is the way data and information is framed that make it utilizable. Business-oriented researchers now tend broadly to reflect the kinds of distinctions Davenport and Prusak make between *data*, 'a set of discrete objective facts about events with record keeping at its heart, *information*, a message which is intended to have an impact ('data that makes a difference') and *knowledge*:

> a fluid mix of framed experience, values, contextual information, and expert insight that provides a framework for evaluating and incorporating new experiences and information. It originates and is applied in the minds of knowers. In organizations, it is often embedded not only in documents or repositories but also in organizational routines, processes, practices, and norms (Davenport and Prusak 1998: 5).

In this sense, then, all workers are 'knowledge workers' and all organizations (even traditional service producers and manufacturers) can be understood as complex knowledge-producing systems and a recognizable part of the Knowledge Economy. However, while there is continuity between the 'old economy' and the new, there is also discontinuity, and this discontinuity is signaled in the shift in the form in which knowledge is produced, stored and used. Established knowledge-building practices have to be reframed if they are to remain relevant and useful.

The everyday conduct of everyday business means that knowledge has to be contextualized and recontextualized as it moves through specific local, physical workplaces if it is to be available for use at remote sites. For some time now, it has been at least theoretically possible to shift data, information and capital around most of the world more or less instantaneously. What this means for production schedules is that they are no longer necessarily governed by conventional concepts like night and day, or a locally available, appropriately skilled and acquiescent workforce. It is always day somewhere in the world and integrated production systems, calling on flexible, trained, globally distributed workforces, can operate 24 hours a day, seven days a week, if the appropriate complex and fluid production, transport and communication systems are functioning as they are supposed to.

Organizations provide both 'data' (in the form of shared data bases, statistics etc) and 'information' (in the form of reports, analyses, production schedules, manuals and standard operating procedures) in an attempt to create comprehensive and context-independent instructions that can be applied equally in any context. Because data and information is encoded in text it moves rapidly across and between contexts, but that is not to say that it is context free. People produce and interpret texts 'in context', and rapid and effective interpretation is critical in global networks of production. Getting the right product or service to the right market 'on time' is the decisive aim in a global economy, but it is rarely in the power of one workplace, or even one global organization, to achieve. More often, it involves people employed in a range of companies and workplaces around the world negotiating with each other to integrate, adjust and adapt their work practices and programs in real time to create an effective supply chain. Supply chains like this are extremely time-sensitive, and very vulnerable to disruption. A transport strike in Melbourne, a fire in a warehouse in Bangladesh or a power failure in Auckland puts everyone's production schedules under pressure and requires people to produce responsive and creative solutions that take account of local conditions as well as global demands. Symbolic-analytic work is the work people often do when things go wrong.

Individual enterprises are not necessarily aware of how these changes to the global economy are playing out in their businesses. Some argue that, far from transforming in response to the demands of networks, corporations are generally still operating as if nothing has changed, 'according to a logic invented at the time of their origin, a century ago' (Zubboff and Maxmin 2002: 3). The basis of this argument is that many US-based businesses at least, continue to be inward looking, focusing on their own internal processes and more efficient ways of doing what they always have done, even in the face of the global demand for connectivity that Castells expounds. While agreeing that ICTs have the potential to transform businesses, Zubboff and Maxmin argue that this potential is for the most part unrealized and that ICTs have been recruited to the cause of reformation rather than revolution:

> The new information technologies have reshaped business processes in many industries. Until now, however, they have been bent to the purposes of the old consumption, according to the principles of the old capitalism. They have continued to do more and to do it more efficiently. If there is some disappointment over what they have yielded this is why (Zubboff and Maxmin 2002: 12–13).

This is not to say that individual companies can exempt themselves from global networks of interaction; they almost never can. Rather, it is to say that they are not managed as if they are integrated into global networks.

Without explicit, systematic Management attention to the ways in which important network connections are established and reestablished within their organizations and between global partner organizations, it is left to individuals and groups of workers employed in the company and in partner companies to improvise connections and retrieve failed connections. They must determine for themselves the ways in which they will recruit technologies to these purposes and integrate them into existing work practices which may be premised on a business model that focuses on internal operations rather than external connections.

In other words, if global knowledge economies are to work even moderately well, people must contextualize workplace texts, taking a critical orientation to data and information, reading and writing it to 'make knowledge' on the run, to collaborate with people they may never have met face to face, who work in environments they can hardly imagine. Individuals and groups of workers in these local workplaces make and use the knowledge they need to meet the demands that the global Knowledge Economy places on them. This kind of knowledge–building relies on print-based and digitized text. Knowledge is textually encoded and knowledge-workers are text makers and interpreters. If organizations are to operate successfully in the Knowledge Economy then they must acknowledge the importance of text-based knowledge to their organization and devise ways of using it to leverage and access other, more established forms of workplace knowledge. The global Knowledge Economy relies on symbolic analytic work, and everyone needs to know how to do it.

TECHNOLOGIES, TEXTS AND KNOWLEDGE WORK

Information and communications technologies

It is tempting to think about the global economy as defined, enabled and constrained by the capabilities of new technologies, shaped by what particular technologies can do. The global economy is, of course, reliant on communications technologies, but it is shaped and constrained, not by the technologies themselves, but by what people can do with them in the specific contexts in which they find themselves. This can be less, more, and different from what technology designers envisage when they imagine idealized information, technical systems, and people, at work.

Contemporary knowledge work is predicated on the ways in which, and the success with which, people integrate ICTs into existing work practices, the ways in which they change existing work practices, and the ways in which existing work practices modify the technologies, or at least influence the ways they are used. The impact of ICTs has challenged conventional understandings of what we mean when we talk about workplaces, not because physical and temporal location is no longer relevant (it is), but because the term no longer seems to capture the extent or nature of the contexts in which people do their work. Increasingly, in my own work and work with colleagues (Farrell 2004; Farrell and Holkner 2004) I am finding it helpful to use two terms, work*place* and work*space*, to help focus on the multiple dimensions in which people do their work.

When I talk about work*places* I am referring to the specific physical location in which a person does the work—the *place*. The idea of the workplace allows us to pay attention to a particular factory floor for instance, with its vast expanse of vertical and horizontal open space, its different generations of machines and the people who operate and repair them, its stores of raw materials and components and the people who order and dispatch them, its computer terminals and the people who communicate on them, its notice boards, its noise, its smells and all the features that define it as 'my workplace' to the people who work in it. In other words, when I talk about workplaces I am focusing on the specific local, spatial, material and temporal conditions of particular workplaces and the people and artifacts that inhabit them. Workplaces are made up of social relationships; they have their own histories and are at least partly products of local and regional historical and political contexts too, so when I talk about workplaces I am thinking of them as embedded in these contexts, shaping them and being shaped by them. The people and the objects that occupy the physical workplaces are not there by accident, and the work practices they develop over the years are at least partly a product of local and regional conditions they experience. Face-to-face communication is a routine feature of most workplaces and face-to-face communication means that a good deal of communication can be implicit, relying on physical gesture, tone of voice, proximity and shared contexts and histories to create shared understandings. Work practices in workplaces have often been unexamined and taken for granted because the people who share the space share narratives about their work, how it is done and why it must be done that way. The work practice can easily become 'naturalized', simply the way things are done and should be done.

The idea of a work*space* is not intended to replace the idea of a work*place*. Increasingly, however, I find it helpful to think of physical work*places* as forming local nodes of a complex network of people, technologies and practices that constitute a potentially globally distributed work*space*. When I talk about work*spaces*, then, I am talking about dynamic, fluid, often transient, working units defined and bounded by regular, routine information communication technology routes (telephone, E-mail, hard-copy print materials, palm-top computers, pagers etc), not bounded by geography and not even by a commercial organization, no matter how extensive and complicated the corporate legal ties might be. I am conscious that I am making a rather arbitrary and pragmatic distinction here. As I discuss in Chapter Two, local workplaces need to be thought about as historically, spatially and temporally embedded *spaces*, too.

In focusing on workspaces as well as workplaces I don't want to suggest that workspaces are entirely new phenomena that have superseded the traditional workplace. I am certainly not denying historical and longstanding webs of production extending well beyond the boundaries of the local firm. It is more the reach and density of connections between workplaces, and the ubiquity of the connections, which seems to need attention. At the local level there is a greater degree of mobility on the part of local working groups than there used to be. People based in the same physical workplace may nonetheless work from cars, airports, coffee shops and home offices, on a regular basis or from time to time. Work is also distributed globally, with particular functions (front office, call-center customer service, back office, finance etc) located in parts of the world selected for the distinctive advantages they offer—a low-cost, English-literate labor force, for instance, or a less organized and more compliant one; proximity to certain raw materials, or modes of transport, cheap land or easy access to lucrative niche markets. The traditional constraints of time zones, physical distance and the political boundaries of nation states still apply, but they no longer apply in the same ways. People employed in different companies communicate by telephone and E-mail, establishing, maintaining and repairing relationships with people they may never have met face to face, operating in physical locations that are far beyond their experience, in physical and cultural contexts they can only speculate about. It is also true of course, that they will use communications technologies to communicate with colleagues at the next desk, in the next room or in the adjacent building.

What is distinctive about work*spaces* is the extent to which the contemporary, globally distributed workspace is created and animated by (and absolutely reliant on) the textual practice of ordinary working people as they go about their everyday working lives. Global production on the scale we see today can only occur because people use whatever information and communications technologies are available to them in their particular workplaces to create the routine and innovative textual practices that bring electronic workspaces to life.

A workspace is defined by routine transfers of data and information, and by the establishment and maintenance of multiple relationships at the individual and organizational level. ICTs change who we relate with, and how we relate, in the workplace. One of the features of ICTs is that, in the absence of regular face-to-face contact, we have to find new ways of representing ourselves. While this is made explicit in contexts like computer-gaming communities, where participants develop complex avatars and other on-line identities, it is less explicit and more complex when identities and relationships are negotiated simultaneously across complex personal (one-to-one) and organizational matrices in hybrid face-to-face and ICT-mediated contexts. Workspaces offer the opportunity for workers to adopt and explore new working identities but these negotiations also refashion established working identities at local workplaces.

There is a further aspect to technology that has a particular impact on the ways people do their work in the global economy. Technologies, especially new technologies, are fragile, and their design often does not take into account the nature of the data and information people actually use, or the conditions under which people do their work. Computers break down, software contains bugs, individual components are incompatible, documents untranslatable across computer languages and programs.

Technologies are part of social systems in workplaces, and it is on the robustness of the social systems that the technologies, and the global networks, rely. As Brown and Duguid argue:

> to achieve the goals technologists themselves set out, it will become increasingly important both to reconceive work and to retool design. In particular, design needs to attend not simply to the frailty of technological systems and the robustness of social systems, but the ways in which social systems often play a key part in making even frail technology robust (Brown and Duguid 2001: 70).

When systems fail at work, people improvise (more or less successfully) to create new knowledge to do the work. When ICT systems fail (as they routinely do, because of design faults in the technology, but also because of power outages, or failures in telephone infrastructure etc) people improvise to bridge the communications gap and keep the global economy operating. The kind of improvising they do is partly concerned with selecting from the new and more established technologies available to them (if the E-mail is down maybe a fax and an SMS to pick it up will do for now), and partly a matter of developing an awareness of the potential social and political impacts of texts so that effective communications protocols and textual practices can be developed. It is important to understand, for instance, that what can be said in a telephone conversation may not equally be said in an E-mail. This is partly because of all the nuances that can be conveyed in synchronous, voice-to-voice communication and partly because of the durability of the written word (whether digital or print). People need to be able to select between technologies and to adapt the text in response to the demands of the particular technology and the anticipated conditions of reception. The conditions of reception can be remarkably different. For example, Arunachalam (1999) describes how developing countries have far less extensive and reliable access to telephones, computers, networks, internet, and bandwidth than economically developed countries. This means that the electronic delivery of large documents and data sets is often unreliable and, even if the infrastructure is available, the appropriate software may not be available to read it. In addition, the impact of business's eating up the available telecommunications infrastructure means that there is less available for local education and community activities. This in turn means that local knowledge is likely to be even more marginalized in global knowledge economies.

Texts, technologies and identities at work

The global economy relies on ICTs, and ICTs create particular kinds of challenges in global workspaces and local workplaces because of the nature of the gaps they create. People need to be able to create texts that take account of the capacities of different kinds of technologies, and their social impact at the site for which they are destined, but they also need to develop a new attitude to textual practices in general.

When I talk about texts here I am talking about instances of written or spoken language; the digitized iTexts (Geisler 2005) that are produced in interactions involving E-mail, mobile phones, the web, personal digital assistants and other technologies, the hard-copy print texts that may be printed documents transferred electronically or by conventional mail, and the conversations that are initiated or mediated by these texts. When I talk about textual practice I am talking about what people do to make and use texts, individually or together.

When people talk about texts at work: memos, tables, E-mails and briefings for instance, they are frequently presented as merely instrumental, as nothing more than more effective or less effective vessels for transporting uninflected information. It can be difficult to understand how such an apparently straightforward process as writing a memo can be so fraught and dangerous. No matter how much we may want or expect them to be, texts at work are no different from any other form of language. The texts people make and use are not, and simply cannot be, neutral forms or practices. Gee makes this point when he says that "While it may suit us to encourage people to believe that language can be neutral in fact it never can be. There is nothing special about using language. Politics is part and parcel of using language" (1999: 2). "Grammar", he says, "simply does not allow us to write from no perspective" (1999: 4). The most important functions of texts are social functions, and this is as true in the factory, the shop, the office and the boardroom as it is in the home, the playground and the football club. The fact that knowledge and information is usually presented as if it were neutral and value free in the workplace does not mean that this is in fact the case. Indeed, the political effects of a text are most powerful when the values that inhere in the text are obscured so that it looks as if the text is reflecting an unambiguous 'reality' and that it has been written 'from no perspective'.

In some important respects, digitized texts cannot be distinguished from traditional texts, they are themselves historically embedded. They have their roots (and, therefore, as Bakhtin reminds us, a good deal of their meanings) in the cultural and cognitive practices of communities established many hundred and even thousands of years ago. When people read or write an E-mail they do not completely put aside their previous (or parallel) understandings of what written and spoken communication can or should convey, and how it should do it. They have understandings of what it is to be polite in writing and in speaking and what can and cannot be appropriately conveyed by E-mail, and while these understandings are evolving, they are nonetheless informed by the communicative communities to which people have belonged.

Gee argues that language has two primary functions: "to scaffold the performance of social activities (whether play, work or both) and to scaffold human affiliation within cultures and social groups and institutions" (Gee 1999: 1). This is a particularly relevant observation when it is made with regard to global workspaces, where the functions of scaffolding social activity and human affiliation are critical and rely more or less exclusively on textual practice. Face-to-face encounters between remotely located colleagues are minimal, and when people engage in joint work practice that practice is textual practice undertaken without the benefit of shared physical contexts. This requires a level of self-reflexiveness not normally required in workplaces where physical and cultural contexts are known and shared, at least to some extent.

Since language is innately social it is also innately political in the sense that using language always entails taking a perspective on our world and our actions, on viewing some things as normal and some as not, some things as acceptable and some as not, some people as 'like us' and some as not. When people are communicating information, collaborating in design, or solving problems at work they are also engaging in an acutely political project, an attempt to regulate what other people will take notice of, and how they will do it, and what they will ignore. They are routinely and often unconsciously making bids for what counts as legitimate working knowledge and who can say so. The success or failure of these bids will have material consequences—it will help determine who makes decisions in a workplace, who gets employed, what they get paid and whose opinion counts. So, when I talk about texts and textual practice at work, even when I do not say so explicitly, I am talking about language as always, and inevitably, social and political practice.

Texts, technologies and ruling relations

So far I've argued that, in the globally distributed, ICT-enabled workspace, work practice has to a significant extent become textual practice, that texts are the contexts in which we do our work, and that these texts, like all texts, are contested sites. Bakhtin explains that:

> The word in language is always half someone else's. It becomes 'one's own' only when the speaker populates it with his own intention, his own accent, when he appropriates the word, adapting it to his own semantic and expressive intention. Prior to this moment of appropriation, the word does not exist in neutral and impersonal language (it is not after all out of a dictionary that a speaker gets his

language!) but rather it exists in other people's mouths, in other people's contexts, serving other people's intentions . . . (1981: 293–294).

All textual practice is in this sense dialogic, reliant on the temporal, spatial and social context of the utterance, and all meaning is transitory and contingent. This is no less true in the globally distributed workspace than it is anywhere else, but it does make global knowledge production, or even global communication, trickier than much of the popular commentary might suggest. The texts people make in global workspaces are a cacophony of potentially conflicting meanings. The words themselves call up multiple historical contexts and competing local intentions and purposes whenever a simple task of text production or interpretation is undertaken, and the social meanings of the particular technologies in which they are embedded and their intersections layer and inflect those meanings. While English is generally the language in which global work gets done, these are forms of English used from many perspectives, inflected with the language and cultures of the producers and receivers, but also with all the historical and contemporary echoes of linguistic imperialism that are attached to English in the contemporary world. From an operational perspective alone, it is amazing that anything ever gets done.

If the global economy is about knowledge production and transmission, and if as educators we wish to support people, communities and organizations as they engage in it, then it is critical that we understand knowledge as embedded and produced within this discursive framework. As Smith argues:

> Knowledge, and hence the possibility of telling the truth and getting it wrong, is always among people in concerted sequences of action, who know how to take up the instructions discourse provides and to find, recognize and affirm, or sometimes fail to find, what discourse tells is there, as well as relying on just such dialogic sequences to settle disputes about what is. Knowledge, thus conceived, is always in time, always in action among people, and always potentiates a world in common as, once again, known in common (1999: 127).

A good deal is at stake in this complex, globally distributed, highly contingent, textually mediated, knowledge-producing work practice. Smith points out that one of things that texts do is to objectify what she calls 'ruling relations':

> Ruling relations form a complex file of co-ordinated activities, based on print and increasingly on computer technologies. They are activities in and in relation to texts, and texts co-ordinate them as relations. Text-mediated relations are the forms in which power is generated and held in contemporary societies. Printed or electronic texts have the generally neglected property of indefinite replicability.

Replicability of identical forms of meaning that can be activated in multiple local settings is fundamental to the ruling relations.

Furthermore:

> The material text creates a join between the local and the particular, and the generalizing and generalizable organization of the ruling relations. It is the materiality of the text itself that connects the local settings at the moment of reading into the non-local relation that it bears (: 79).

From this perspective, the textual practices of global workspaces are not only knowledge-producing practices, they are the practices in which power is 'generated and held'. When people engage in the textual practices of 'making knowledge common' in local and globally distributed workplaces, they are participating, with greater or lesser effect, in the microprocesses of the production of global power relations. Their capacities to engage knowledgably in the textual practices of the workplace, and to develop and promote new (possibly competing) textual practices, will have a material effect on their working lives.

So much for the macroprocesses of economic globalization. My focus for the rest of this book is on the microprocesses of economic globalization, the things particular people do to make globalization happen. In Chapter Two I take up the challenge of thinking about the Exmouth Plant and the Harbor Mill as simultaneously workplaces and workspaces and what this means for the people who work there. In Chapter Three I take up the question of 'knowledge' and how it is made and understood. In Chapter Four I focus on a particular incident of knowledge production as it happens between colleagues but shaped and constrained by global regulations. Chapter Five looks at knowledge production in 'cyberplace'. In Chapter Six I look at the impact of all this on working identities, and in Chapter Seven I focus, at last, on the role of workplace education, and workplace educators, in the processes of global-knowledge production.

CHAPTER TWO

Workplaces/workspaces

> It is recognized that we always act (and think) locally, but our actions and thoughts are simultaneously urban [or rural], regional, national and global in scope, affecting and being affected by, if only in the smallest way, the entire hierarchy of spatial scales in which our lives are embedded.
>
> (SOJA)

> Electronic business and new organisational forms explode the classical boundaries of the firm with regard to time and space. Workplaces, business processes and even whole organisations are no longer bound to locations the way they used to be.
>
> (MOSLEIN)

> In theory, more open global markets and faster transportation and communication should diminish the role of location in competition . . . But if location matters less, why, then, is it true that the odds of finding a world-class mutual-fund company in Boston are much higher than in most any other place?
>
> (PORTER)

INTRODUCTION

Despite enthusiastic early claims, communications technologies have not resulted in the paperless office, nor have they produced an environment where work can be done 'anytime and anyplace' (O'Hara-Devereaux, M. and R. Johansen 1994: 199). Time and place still provide the contextual set points around which people go about their working lives. But to acknowledge that time and place matter is not to say that they matter in the same way. Navigating time and place has become a more complex process as people juggle multiple time zones and multiple places (both local and global) in the everyday conduct of their work.

While electronic communication has meant that "workplaces, businesses and even whole organizations, are no longer bound to locations the ways they used to be (Moslein 2001: 1)", they are still often bound to locations, although with less immediately obvious ties. Boston remains the home of major finance houses, Silicon Valley attracts software development companies. Physical spaces offer specific attractions to particular forms of activity and proximity, and the possibility of face-to-face connection and incidental trading of information, gossip and opinion still offer competitive advantage. Electronic communication has not neutralized physical location, but it has made it more complex and more difficult to see and understand. It has required the reconceptualization of location to take account of multiple, intersecting physical locations, some immediate and some remote.

When we think about the contemporary workplace and about contemporary work practice, then, we need to think about it as existing simultaneously in an immediate built environment (an office, or a farm, or a factory), in a geographical location (in a city or a town or a village) in a region with its own local economy, a nation with its history, and its internal political imperatives, and within a global environment. Globalization is exactly like other social phenomena in that it is produced, moment by moment, in the microprocesses of people's everyday lives, it "is produced/accomplished by people 'at work', that is active, thinking, intending, feeling in the actual local setting of their living" (Smith: 75). People engage in the global economy from somewhere. Global workspaces cannot be understood independently of local, physical workplaces, but neither can they be understood simply as the aggregate of local workplaces, joined together by neutral communication systems. All work practice is geographically, socially, culturally and temporally situated, although it is not always easy to see how.

The challenge of making knowledge common is a challenge faced by ordinary people as they go about their routine work in their specific local settings, day by day. The objects, resources, practices and relationships of these individual local workplaces make up global workspaces too. A great deal is assumed in local workplaces, it has to be. The 'taken for grantedness' of local workplaces and local work practices allows people to get work done efficiently, communicating elliptically and accepting, perhaps not even noticing, the ways objects, values and actions are integrated into daily working lives. The taken for grantedness of local settings poses a problem for global workspaces. What is taken for granted is invisible to outsiders, but it is also invisible to insiders, perhaps even more so. It is difficult to become conscious of how our practices are shaped by our histories, our physical location, etc. While outsiders may be able to see features of our environment that have become invisible to us over years of familiarity, they do not know what meanings we invest in them, or how they are related to each other. Once we begin to understand the unstated assumptions about place, time and practice that underpin local work practices and working relationships and the ideas and values that are implicit in them, we are in a stronger position to understand what is involved in making and sharing knowledge across communication networks.

Texts join up local workplaces, but they do not create a neutral, context-free 'cyber-place', a place in which local contexts are neutralized and people can assume context-free identities with which to relate with others. In some respects, cyber-place is just like any other place. It is replete with the meanings invested in overlapping physical places and ambiguous and contradictory working and social identities; it is made up of multiple and conflicting discourses and multiple subject positions. There is more context, not less context, to understand. Global workspaces are textual phenomena, but it is not clear how texts are inserted into specific physical locations, how they change work practices and how existing work practices change them.

In short, to understand the textual production of knowledge at work we have to understand where it happens. This is not just a matter of describing the physical layout of a workplace, it involves coming to an understanding of why that particular geographical location and not another, why it is populated by people of a certain age, nationality, gender and not others, why (and how) they use certain objects to make and use knowledge, and not others.

Soja's framework for understanding cities in terms of their physical characteristics, the social practices and ideational frameworks that produce those characteristics, and the broader historical and political patterning of the local, regional and national landscapes, provides a helpful framework for attending to the taken-for-granted features of local workspaces. I want to use that framework here to bring one local workplace into focus as a physical and geographic entity before looking at the way it operates as a node in a global network of production, a global 'workspace'.

Soja is interested in understanding contemporary cities in a world marked by the accelerating global flow of people, ideas and resources. He starts from the position that the development of any city is linked to the dominant mode of production. The industrial metropolis, for instance, is inextricably linked to the specific needs and processes of industrialisation. The first factories transplanted crafts like weaving to places where energy was cheaper (initially close to a reliable source of water) and where supplies of less organized (in the industrial sense) workers were plentiful. Cities are spatial forms and they arise from social processes (: 99) and the pressures of the current environment of 'fast capitalism' are reflected in contemporary urban development.

Soja uses the idea of 'city space' to focus on cities as actual spaces nested within, and profoundly influenced by, localities, regions, nations and global environments. He understands a city space to consist of three spaces. The 'first space' is the physical and empirical patterning of space, the perceived space, the things in a space. The 'second space' is the ideational field—the conceived space (the way we make sense of the space) and the thoughts we have about the space. The 'third space' is the way the material and ideational dimensions of the space come together, "the real and imagined, actual and virtual locus of structures individual and collective experience and agency" (: 11). Like cities, individual workplaces and workspaces can also be thought about as 'third spaces', physical and ideational instantiations of particular modes of production, nested in the broader social networks of the cities or regions, and the networks of cities, to which they belong. I find this a useful way to think about workplaces and workspaces, and to begin to understand the complex ways in which people, place, ideas and practice interact and are interrelated in electronic workspaces.

In this chapter I want to analyze one company, Australian Fabric Manufacturers (AFM), from the perspective of the 'third space'—the physical and ideational space brought together. People inhabit workplaces, and they bring with them their individual biographies. When they talk and write at work, when they make texts, they invest these texts with intentions formed in these biographies. When people collaborate in knowledge-making textual practices they do so from their own biographical positions and in dialogic relation with each other and with their local and global contexts. In the course of this book I will introduce a number of people who work at AFM. For now I would like to begin to introduce four of them, each critical to the knowledge work of the company and in their own ways indicative of the ways in which historical and political processes shape who gets to work where, and why.

Bill is a Warping Shed Supervisor at the plant, an immigrant from England who came to Australia under a government-funded skilled trades immigration program in the 1960s. He trained as a Machine Lacemaker in the UK, leaving school at the earliest opportunity to undertake a rigorous apprenticeship, and has developed those skills over a long career. Mary is the Mending Room Supervisor at the Harbor Mill. She has lived at Harborside all her life, leaving school at the earliest opportunity to work at the Harbor Mill with her mother and sister, resigning to raise her family and returning when her children were all in secondary school. Baz is a Production Supervisor at the plant, training as a textile technician through an apprenticeship at AFM and moving into middle management after twenty years on The Floor. Grace has lived at Harborside all her life, leaving school to train as a Chef. She worked in hospitality for five years but tired of the long hours. She was recruited to the Harbor Mill where her cousin works and her mother had worked before she died. Grace trained as a Weaver, then a Warper, and is now Warping Shed Supervisor at the Harbor Mill although not yet thirty years old.

Contemporary textile workers like Grace and Mary and Baz and Bill, and contemporary textile production itself, are shaped by the historical processes that produced the textile industry and by the social and economic processes that are influencing it now. The ways they engage with the global economy, the ways they make and use texts, and the ways they understand what counts as knowledge and who can say so, all rely to a significant extent on historical and physical contexts and the power relations that are worked out in them.

THE HISTORICAL PRODUCTION OF TEXTILE PRODUCTION

Australian Fabric Manufacturers is a well-established Australian company, part of a number of global networks, and largely concerned with producing sophisticated technical fabrics, specifically upholstery fabric for the global automotive market. It is, therefore, situated at the intersection of two major industries: the textile industry, the development of which was so significant in the industrial revolution, and the automotive industry which is emblematic of the current revolution in globally distributed production. When Mary and Baz and Bill and Grace warp and weave and mend at AFM they do so in workplaces that bear the traces of 200 years of factory-based textile production but which are also fundamentally shaped and changed by the demands of producing a quintessentially 'global product'—the world car.

Until the 1800s the production of wool or cotton cloth in European countries happened at home. The work was generally divided between the members of one family along gendered lines, the women and girls cleaning the fleeces and carding and spinning the yarn (hence the term spinsters). The looms were usually located on the upper floor of the house, where the windows were large and the day light was good. The men of the household were the weavers; the looms required simultaneous hand and foot movements which demanded agility, strength and endurance, so weaving was considered physically hard work and consequently the province of men. Working in the home in this way meant that people had a good deal of control over how they spent their time, and housework and work for money were not easily distinguished. Everyone in the family made a contribution to the different kinds of work. The rhythms of the working day, and the year, were partly dictated by the demands of cloth production but partly also dictated by the seasons, and by various tasks that needed to be performed to keep the household going.

Textile production was in the vanguard of the Industrial Revolution. The comparatively large and expensive machines demanded many more workers than a single household could provide if the owners were to recoup their investment, and they required specific conditions that were most easily and economically met in a factory setting. Textile work, which had been distributed across a number of homes and families in a village or town, came to be concentrated in one place. Textile workers had to travel to work, sometimes living and working away from home for long periods. The machines in the factories were expensive to buy and factory owners wanted the machines to operate for as long as possible to maximize production, so it was the factory owners, rather than the spinners and the warpers and weavers and the winders, who determined when and for how long people would work. Textile work remained gendered, but the distribution of the jobs changed. The machines reduced the amount of physical effort required to operate a loom and weaving became the province of unmarried women, who could be spared from home and live in dormitories. They could be paid less than men but still earn more than they would as farm laborers, and live with greater independence. Warping (the setting up of the yarn to provide the framework within which the cloth was woven) was viewed as skilled work, hard work, and men's work. Some of the work in factories required workers who could maneuvre between and under machines. Small stature became important, and the workhouses provided a plentiful supply of children, so child labor was acceptable, and children could be paid very little.

So, the change in the physical workplace involved a powerfully significant change in the ways that work was organized and understood. Where work had been flexible in both time and space when it was located in the home, in the factory work became fixed in both time and space. It became important to count working hours, to divide the day according to work shifts, to be paid according to hours worked, to regulate the hours worked rather than simply work till the job was done—it was never done. Craft Guilds, which had previously been mostly concerned with inducting apprentices into the discourses and practices of their trades, began to take a critical role in collective bargaining over pay, hours and other working conditions, so membership of guilds and unions was associated with both craft knowledge and expertise and a specific, adversarial relationship with factory owners and their managers (Simpson 1999). Weaving, once the province of men, became the province of women, and tying in, warping and other more technical dimensions of textile production became men's work and remained so, by convention, until relatively recently.

As Soja argues, spatial forms, like the textile factory, arise from social processes, and they operate to produce and reproduce those processes (: 99). Traces of the distant history of textile production, in another part of the world, can be seen in the physical and temporal organization of work at AFM. They can also be seen in the gendered nature of the distribution of work and in the way people feel about it. For Foucault, space is where the discourses about power and knowledge are transformed into relationships of power (: 50). A workplace is precisely that kind of space, and the textual practices of work take up just those kinds of struggles. Bill, Baz, Grace and Mary engage in those struggles as they make and use knowledge, and they do so with the assumptions, resources and historical and physical constraints of the places in which they work. Relationships of power in workplaces are often worked through in terms of what counts as knowledge and who can say so, and history plays a part in shaping what people take for granted and what they believe is, or can get recognized as being, up for negotiation.

AFM

The story so far . . .

The Head Office and the original site of AFM is the Exmouth Plant, located in what used to be the heartland of textile production in the inner-northern industrial suburbs of Melbourne, the second largest city in Australia. The textile clothing and footwear industry, like that of the rest of the Western world, has been radically restructured in response to the reduction of tariffs and the opening of markets to imported textiles, notably from Asia. AFM is, as a consequence, no longer situated within a concentration of textile production. While its physical location remains the same, it is now surrounded by a mix of light-industrial buildings and new housing developments designed to appeal to people who work in the city but cannot afford to live there, or to live in the inner suburbs traditionally designated as residential. So the AFM Head Office and plant has something of the feeling of an outpost and monument to a previous era. As the city of Melbourne becomes larger, inner-urban land once used for industrial purposes is being claimed for residential use and the cost of land for industrial use is now considered prohibitive. The land on which the Exmouth Plant sits represents a significant capital investment for the company, and it is unlikely that the metropolitan plant will expand.

The Exmouth Plant was established in the early 1930s. Its original purpose was the production of warp knitting/single-loom fabrics and particularly the then-revolutionary synthetic fabric, rayon. This fabric was used in the manufacture of lingerie, mainly for the production of satin-striped underwear. As the business prospered the owners invested in a new plant and technology to produce a wider range of lingerie. The new plant made it possible to expand the lingerie line. Notably, it allowed for the production of the new synthetic fabric velour, which was used for the manufacture of dressing gowns. The subsequent success of the company, when all around it textile companies were closing their doors, rests to a significant extent on its largely serendipitous decision to tool its plant to produce velour dressing gowns. In the late 1960s AFM acquired a company which had the technology and expertise required for dying and finishing. It was no longer necessary for the company to rely on outsourcing these functions, it could control the whole process of fabric production in-house.

Meanwhile, in an entirely unrelated industry, fashions were also changing. After the Second World War the automotive industry expanded to become the largest industrial sector in the world. Until the late 1960s cars where upholstered in leather, at the top end of the range, and vinyl, at the mid and lower levels. In the early 1970s this long-standing practice changed. Vinyl was no longer being used for seat coverings and fabric, especially velour fabric, was taking its place. At that time it was the practice of the automotive industry to source most of its components, including its upholstery fabric, close to its assembly plants. AFM was located close to a number of vehicle assembly plants which supplied the Australian market. The plant was already producing velour fabric, for dressing gowns, so it did not need the time or the finance to significantly retool. It took advantage of the opportunity to diversify and then to shift its production from the shrinking lingerie market to the expanding automotive market. It was a successful move, and AFM became the first Australian company to supply automotive upholstery fabric to the major international automotive manufacturers. Using the technology and expertise accumulated in the development of velour for apparel, AFM developed and produced an increasing range of automotive velour fabrics.

AFM's diversification into automotive upholstery proved to be strategic. In the late 1980s, as trade barriers were removed and the textile industry in Australia and all over the world was rationalized, AFM acquired a number of other companies in the immediate area and in other textile-producing locations. While the real estate associated with these companies was generally sold off, the Harbor Mill and one plant in a neighbouring suburb, was retained. The later was generally concerned with the dispatch of fabrics overseas, and it was the space rather than the plant that was attractive. AFM had no real plans for the site and made no investment in it (at their lowest point employees had to petition for a first aid kit), and the site was ultimately sold and the workforce redeployed to Exmouth. The Australian operation of AFM is currently distributed between the Exmouth Plant and the Harbor Mill.

In the 1990s AFM was again experiencing the need for new technology, but as a private company it did not have the capital required to invest in a new plant. Instead it entered a joint venture with a Japanese company which had state-of-the-art technology and a different and more contemporary design capability. The management styles of the two companies were, in the end, incompatible, and AFM ended the relationship in the late 1990s. By now, however, the company was keen to enter the Asian market, as its website proclaims, to 'expand its niche markets globally with an emphasis on the Asia Pacific'. Through various changes of ownership, but with the original private owners still maintaining significant control, it attempted another joint venture in South Africa and joined with local and European partners to move into the Indian market as a local manufacturer, building a new plant in India with specific technology. From the bemused perspective of many employees, the company went from competing with Asia to 'being Asian', seemingly overnight.

Over the past seventy years AFM has moved from being a single-site domestic manufacturer of synthetic lingerie products to a multisite manufacturer of high-technology industrial fabric, trading largely in the global automotive sector. One of the consequences of trading in a global market has been the requirement to achieve international quality standard certification. AFM has two significant international standards certifications—ISO 14001, which is concerned with international environmental standards, and QS 9000—the first which is concerned with automotive standards. It is the first textile company in Australia to be awarded quality standards within the automotive standards framework. In 2001 it was acknowledged for 'Excellence' by a major multinational automotive assembler for its materials, services and consumerables supply. Most of AFM's training effort is driven by the demands of various international quality frameworks. Some, like those mentioned above, are generic and others are highly specific, generated by several different global automotive companies. AFM relies on training to meet the formal requirements of these quality frameworks, and satisfying these frameworks allows the company trade in a global marketplace dominated by the automotive sector. AFM is not only 'becoming Asian', it is relinquishing some its close ties to the textile industry, and becoming 'automotive'.

The Exmouth Plant

The Australian operation of AFM exists at two physical locations. The Exmouth Plant is a two-story building consisting of a lower floor and an upper floor. The lower floor is the factory floor, and is generally referred to as 'The Floor' (he's gone onto The Floor, she's down on The Floor). This is where fabric is produced, checked, finished, and dispatched, stock is controlled and orders are scheduled. The first thing any first-time visitor will notice is the noise. At the entrance there is a basket of earplugs and visitors are advised to take them and wear them. Most of the older and more experienced workers on the floor disdain them. Machines dominate the workspace, dwarfing the workers who operate them. Some of the machines are very new, notably a vast, and extremely expensive, warping machine purchased to produce the warp for the new generation of synthetic fabrics. The other machines reflect state-of-the-art purchases at different periods in the plant history. At any time several machines are being repaired by technicians or are awaiting parts. Some machines are older, more likely to break down and more complicated to fix. Few of the younger technicians know how to repair them and the company relies (with growing unease) on Bill, a Warping Shed Supervisor perilously close to retirement age, to keep them going.

People don't talk much on the The Floor. The noise of machines drowns out any incidental conversations, and people only make the considerable effort to talk and to listen if there is something important to say. Notes, taped to machines and pinned to notice boards, convey cryptic information about the conditions of the machines or changes to scheduling. Almost everyone on The Floor is a middle-aged or older Anglo-Australian male. Two male Vietnamese workers have recently been employed. They keep to themselves. The Productions Manager, the Systems Manager and two or three staff (one a young woman) work in a large room immediately off the factory floor. The room is papered with production schedules, and the workers, for the most part, work at their computers. Next door is a small training room that is too small and too noisy to be routinely used for training. Where training occurs it occurs 'on-line' (on the production line, that is, not on a computer) with the trainers working side by side with the workers as they do their work, or in training rooms in other parts of building. There are computer terminals on The Floor, used by supervisors to check production schedules, availability of yarn and other supplies and specifications.

Upstairs, on the Fluffy Floor (it is carpeted) it is, remarkably, quiet. Modular partitions divide the floor space into offices and meeting rooms and people sit at their individual computers and do their work. Various formal meetings are held to coordinate activities and the interests and concerns of The Floor are conveyed by the Production Supervisor or the Systems Manager. Most of the workers on The Floor avoid the Fluffy Floor, even arriving for work and leaving by a less convenient entrance rather than walking though the alien office space.

The Harbor Mill

The Harbor Mill is housed in a vast single story building on the outskirts of a provincial city about 100 kilometers from the Exmouth Plant. The most striking thing a visitor might notice on arrival is that, although there is a reception desk immediately inside the front door, it is unattended. Someone unfamiliar with how things are done at the Harbor Mill might wait for quite a long time before a passerby tells them to open the door to the left and bring their inquiry to one of the office staff. If they did that they would find themselves in a partially partitioned office space of eight to ten desks, some behind glass windows but most in an open space. Some desks are clearly not in regular use, with books and papers covering the desk in high piles. At other desks women and men work on computers and talk on phones. Any one of them might answer your query. At the end of the office space is a sliding door to the factory floor. Muffled noise from the factory seeps through the door when it is closed, and it is often left open. There is regular traffic through the door, office to factory and factory to office. As people pass through they call each other by their first names and continue running jokes about the Harbor Mill football team. The factory itself is divided into traditional spaces, the Weaving Shed, the Warping Shed, Finishing and Dispatch. Overall the workforce is Anglo-Australian, more or less equally male and female, and comparatively young. Perhaps a third of the workers are under thirty. The Weaving Shed and the Warping Shed are dominated by large, relatively new, and noisy, machines. Baskets contain earplugs and about half the workers use them. Supervisors work in small offices that are no more than partitioned floor space from the warping and weaving sheds. The Mending Room is a place apart. It sits behind a heavy closed door near the Finishing Department, and from the outside there is no indication of what goes on behind it. Anyone who pushes open the door is greeted by a domestic interior. A radio plays. If it is December there are Xmas decorations, cards and a small Xmas tree. Some workstations are decorated with pictures of children and grandchildren. Here the women are Anglo-Australian, and middle aged or older. They sit easily at their mending tables, pulling fabric down on to the table and examining it, pulling out threads and weaving in new threads with domestic sewing needles. Under the tables are boxes of colored pencils, used for coloring in incorrect threads when the pressure of time is too great to allow correct mending. Mary, the Mending Room Supervisor, sits at a mending table, at the front of the room, working on her own pieces of fabric, and often leaving her workstation to help others. She is also responsible for keeping track of the bolts of fabric. She does this in the traditional way, making entries in an exercise book, and the modern way, entering data into a computer program located in the Mending Room. When she works at the computer terminal Mary must turn her back on her colleagues, and absent herself from the gossip, advice and chat that flows across the room without a break.

In the large Warping Shed on the other side of the Mending Room door, at the beginning of each shift, the young Supervisors meet with their work teams to discuss problems with machines or yarn supplies, timelines, new and tricky patterns. Generally, they call out over the noise of the machines but sometimes, if the difficulties are complex, or emotions are heated, they stop some of the machines so that they can all 'hear themselves think'. By each machine there is a Day Book, the means by which people on one shift communicate with people on another, recording problems with a particular machine, reasons why they are behind (or, sometimes, ahead of) schedule, and any solutions they have discovered. Grace, the Warping Shed Supervisor, is younger than almost all the warpers, and a woman as well. She is the first woman Warping Shed Supervisor.

Workspace, city space, region, nation, globe

Soja encourages us to think about any social action (like working or learning) as being embedded "if only in the smallest way, in the entire hierarchy of spatial scales in which our lives are embedded" (: 200). What this means for Bill and Baz and Grace and Mary is that, as they do their work, their actions are embedded in their workplace and its particular history, their city, region and nation and the global environment in which they operate in ways that are likely invisible to them. Exmouth and the Harbor Mill are both AFM workplaces, but they are distinctive environments, with distinctive histories, and the work practices and social relationships reflect and enact those histories and locations.

The Exmouth Plant, for instance, is a two-story building located in inner-urban Melbourne. It is situated in what used to be the heartland of the textile industry in Melbourne, close to complementary plants, close to the city and close to the airport. The two-story construction reflects the fact that by the time the company needed to expand, land was already getting expensive in inner

Melbourne, as people sought affordable housing close to the city. The rigid demarcation of the two stories, with an almost impermeable barrier between the lower story devoted to production and the upper story to management, reflects a rigid traditional demarcation between Tradesmen and Managers. Virtually all the Textile Workers on The Floor left school at the first opportunity and learned their trade as apprentices, or were employed as unskilled workers and learned informally, by sitting next to more experienced workers and learning 'on the job'. Bill, the Warping Room Supervisor who knows how to fix machines, for instance, was apprenticed as a Machine Lacemaker in the south of England in the early 1950s, and is a product of a broader and more rigid indentured training system than generally pertained in Australia. People working in the offices upstairs, on the Fluffy Floor, by contrast, have finished formal schooling and have generally completed an additional qualification at university or at a technical and further education college.

The Exmouth Plant is situated in the midst of a large potential workforce, including many people of immigrant background, but attempts to recruit, and particularly to retain them, as workers at the plant have not been successful. Notably unsuccessful have been attempts to recruit younger workers, the attractions of cleaner work, and work in the city, seems to be too powerful. The Exmouth Plant is equipped with generations of machines, some only approximately fit for purpose, which mark the gradual transition of the company from lingerie manufacture to technical and automotive upholstery. The smooth functioning of the plant relies on a few people (Bill and Baz in particular) who can identify the difficulties with the machines (the yarn is from a new source, and irregular or too slippery, for instance) and devise ways to fix or get around these problems. The history and geography of the Exmouth Plant, and of the textile industry more generally, is written in its external and internal architecture, in the functional organization of its space, in the objects that occupy the space, in the social and cultural characteristics of the people who work within it and in the ways they do their work.

The Exmouth Plant and the Harbor Mill are two sites of the same textile manufacturer and yet distinctly different workplaces. The Harbor Mill is a vast single storey building on the outskirts of the provincial coastal city of Harborside. Land has been cheap in Harborside for many years and a simple single-story construction is the most inexpensive way to build here. The building is rundown, depressed and depressing, reflecting a long history of a depressed manufacturing sector in the area and the frequent changes of ownership the Harbor Mill has experienced over the previous ten years. The unattended reception desk marks the recent shift of front office functions to the Head Office at the Exmouth Plant. While there is a dividing wall between the office and the production areas, staff from all parts of the building enter and leave through the office, which is the most convenient route to the car park. The easy passage of workers from factory to office and back is facilitated by the building design but also by the absence of senior management, who all now work at the Exmouth Plant. Only those office functions directly associated with the Harbor Mill are now carried out on the premises, and they are checked and approved at the Exmouth Plant, a process made possible by extensive use of E-mail and other electronic communication.

The young people employed at the Harbor Mill reflect a number of regional conditions. Five or so years ago unemployment in the region was twice the state average and the range of employment options for young people was narrow. While unemployment levels are now at or below the state average, this reflects growth and investment in existing industry sectors, including industrial textiles; not an increase in the range of industries. Harborside is an old town, with a significant history, and it provokes strong family allegiance. Many young people want to live and work in Harborside, close to their homes and families. The strength of local family ties is evident in employment patterns. Many of the employees at the Harbor Mill are related to each other or part of outside friendship groups, especially sporting clubs and church groups. When the local Trades and Labor Council made a submission to the Productivity Commission it did so on behalf, not just of Harborside workers, but on behalf of their families and communities as well.

Historically, Harborside has been a manufacturing center since industrialization came to Australia. It presents itself, with some justification, as the national hub of textile manufacturing. Its two main industries are textile production and automotive assembly and components manufacturing. Harborside is a 'node and nucleation', to use Soja's term, in a global web of production, but the dynamic is between industrial textiles, automotive component manufacture and assembly, and industry-sector research and targeted training. A large university with its main campus in Harborside has recently opened a technological precinct in the city and the local technical and further education college specializes in delivering a range of training programs in these industries. Whereas the Exmouth Plant was formerly located at the center of one sort of cluster, the Harbor Mill is now located at the center of a kind of cluster that matters. The Harborside region is also benefiting from being electorally sensitive. Recently, state elections have been won and lost in regional Victoria and the state government has recently upgraded the main highway to Melbourne and promised to upgrade the rail link. While this has advantages for individuals wishing to travel to and from Melbourne regularly it benefits industry most by decreasing the transport costs.

Physical place matters in the production of working knowledge. It provides the historical and physical contexts—the setting and organization, the objects and artifacts—around which people develop their understandings of what work is and how you do it, and it defines and constrains what topics can be discussed, what can be said about them, who can speak and who can write, and what they can say. It is, of course, also important in determining how texts are read, understood and acted upon or ignored, and what happens as a consequence.

However, as I argued in Chapter One, physical place is not the only space that matters. Industrial spaces like the Exmouth Plant and the Harbor Mill are embedded in a network of electronic connection. Employees routinely communicate with customers, suppliers all over the world and with state and federal bodies which support and regulate industrial and trade activity. They are in regular and routine contact with other elements of supply chains of which they are part, once again potentially located all over the world, and with global regulatory bodies. Taken together, these and other communications connections constitute textually mediated electronic workspaces in which geographically local work practice is reconfigured and reinterpreted and global work practices are produced. It is not always easy to understand the ubiquity of these connections, or the ways in which local work practice contributes to, and is shaped by, global supply chains made possible by ICT enabled connections. In the section which follows I want to give a flavor of what kind of world AFM entered when it shifted production from dressing gowns to car seat covers.

The world car: a global product

In the 1950s and '60s the major challenge for car assembly was the fitting together of large, heavy and unwieldy pieces of metal. These pieces were expensive to transport, so car manufacturers located assembly plants in or near major markets, and they sourced components from domestic suppliers, who developed the technology they needed domestically, taking account of locally available materials, local skill bases, local work and industrial practices and the ways in which the local community organized itself more broadly. In this sense, although the source design of the car may have been developed by a global company, it was 'local' in that many of the components were locally designed and made, using and developing local knowledge. It was also 'local' in the sense that specific adaptations were often made to the source design to take account of local preferences. These preferences could affect fittings and trim, more suitable suspension, refinements to account for the effects of heat and dust, and compliance with local regulations about safety, emissions, and right-hand or left-hand drive.

Until the 1980s, new models of cars were developed over an eight to ten year period. This allowed time for local plants and local components manufacturers to develop techniques, technologies and materials, and to train the local workforce in their use, before a new source design was introduced. While the assembly plants were generally wholly owned subsidiaries of global companies, they were often managed by local people who had some autonomy as far as local decision making went. The 'localness' of the operations was often stressed, by company advertising and by governments keen to justify special incentive schemes to attract foreign-business operations. Many local communities relied heavily, almost exclusively, on the automotive industry. The assembly plant itself employed many workers, but the plant attracted components manufacturers and allied industries so that whole families were employed in the industry, and the ties between the various manufacturers were social and familial as well as commercial. Knowledge and skill were shared around the kitchen table and at the club and the pub as much as they were shared on the shop floor or the production line.

By the 1990s the 'world car' had emerged, and the local automotive industry, and the local communities and knowledge bases it supported, were changed forever. For decades the technology involved in the manufacturing of cars was relatively straightforward but increasingly the production of a new car, slowly through the '70s and then more rapidly, came to involve the integration of very different kinds of new technologies. Cars became primarily electronic rather than mechanical, and they were built of new light (and more easily and inexpensively transportable) materials–plastics and synthetic ceramics, aluminium and lightweight steels. The physical weight of the components became less of an issue, and the requirement of 'state-of-the-art' materials and components became far more critical.

As cars became more complicated, assemblers focussed on the design of cars and the integration of diverse and sophisticated new technologies; they left the actual development of the new technologies to trusted, globally oriented component manufacturers who were in a financial position to invest heavily in highly targeted forms of research and development. The only way that local component manufacturers could supply local assembly plants was if they had a technology licence from one of these global components companies. They could no longer develop and use their own technology because car assemblers would not use it.

Even then, car assemblers were wary of companies operating under licence because of the time delay in getting updated information to licensees. As I noted in Chapter One, 'time to market' is critical in global economies, and time has become the critical factor in global automotive production. In the 1980s a new model car took eight to ten years to develop. By the year 2000 the development time was down to thirty months. Cars are now designed in close collaboration with suppliers who are developing materials and technologies during the design processes. They do not wait until after the design has been completed to develop products to specifications, but may sometimes even develop materials which are incorporated in, and therefore change, the overall design during the design process. Design is an ongoing, collaborative process, iterative and contingent on every innovation. Designers need not only to produce innovative new technologies, they need also to be able to adapt them to accommodate someone else's good design idea.

Another factor in the development of the global car is that cars are no longer designed for single markets, although they may be designed for specific demographics. At the top end of the market a fundamental requirement is that cars are more or less identical wherever they are produced, so the car is truly 'global'. At the bottom end of the market cars are designed for cross national markets. Even where adaptations for local conditions are made, the adaptations are not necessarily a local activity. For instance, one South African plant adapts cars for operation in severe heat and dust, and exports those cars to the Middle East and Latin America (Barnes and Kaplinsky 2000).

The 'world car' puts particular stresses on workers located at remote sites. As the product is always in process, workers must be constantly updated on changes and developments in processes, practices and materials. But, as the product is generally designed and developed elsewhere, workers have no firsthand involvement in the design process, and they are not likely to know anyone who has. As the product is standardized across a global network of manufacturing, and as modular design means that several components are assembled in one module before it is shipped, it is difficult to get an overall sense of how the design works. Workers need to be continuously educated, but it is more and more difficult to learn, at least in traditional ways.

When, in the 1960s, AFM moved to produce automotive fabric, these were the processes into which it inserted itself. As the pressures of accelerating time and diminishing space become more acute, every member of AFM finds their local work practice changing, involving, more and more, comprehensive regulation of the most local and intimate work practices. Increasingly, all employees communicate with, and adapt to, people and organizations on the other side of the world. It is not surprising, then, that they are bemused when they are asked what sector they work in, what sites compromise the company, or who makes the call that a batch of fabric does not meet quality standards.

AFM: a third space 'under construction'

Third spaces bring together material and ideational spaces in a complex and dynamic matrix of meanings. The material spaces of the Exmouth Plant and the Harbor Mill are joined together by material texts, and those texts are part of the materiality of AFM. Like the material space, the ideational space is under construction. The traditional-gendered distribution of work, for instance, is being undermined–Grace is not just a warper but a Warping Shed Supervisor. The shift in production, from lingerie to industrial textiles, has caused AFM to be gradually integrated into the production of the global car and so the very construction of employees' work as 'textile work' is under challenge. As the company becomes involved in the production of the global car it becomes necessary for it to engage with specific global teams; to collaborate in the construction of a global workspace. The way it engages with these teams is through textual practice, and it is formally regulated by texts like Standards Frameworks and Quality Frameworks as well as more immediately regulated by electronic and hard-copy memos, informal E-mails and written notes, etc.

Caught up in the same web of textual regulation, indeed, involved in the production of the same cars, is the hydraulic manufacturing plant in Cape Town described by Sholtz and Prinsloo (2001). While subject to the same textual regulation as AFM, this factory is indelibly marked by the specific history of apartheid. Machine operators and team leaders (exclusively 'Coloureds/Cape Coloureds') speak Afrikaans to each other, and many do not speak, much less read or write, English at all. Most have had access to very little schooling. English is the language of Management. Team leaders speak and write English and so can create the local join between Management (mostly 'Whites') and worker and the global join between the company and the global workspace that makes the global car possible, and the Cape Town hydraulic manufacturing company viable. When people at AFM engage in the construction of a global workspace, to solve a production problem or to renegotiate time lines for instance, they are engaging with a very different physical and ideational workplace, with profoundly different power relationships. Words, and even the use of the English language itself, will invoke different meanings, carry different resonances, and have different consequences, but all this will be invisible, or only very vaguely perceived, by other participants in the process.

The material texts that join up the Harbor Mill, the Exmouth Plant and the Cape Town hydraulics factory are not neutral. They are inserted into these physical locations as objects, and as objects they have resonances—perhaps that textual work is not 'real' work (or not men's work, anyway), perhaps that if I use English I shift my working identity from worker to Management (and, perhaps, 'sides', in a larger, national, socio-political struggle). The texts themselves create workspaces which compete with physical workplaces for primacy, but they also change physical workplaces by changing physical work practices (although not always in the ways they are intended to by the textwriters). Noone is isolated from the various global workspaces that AFM inhabits, but not everyone is aware of the contribution they make to the construction of these workspaces, nor are they necessarily aware of precisely how their local work practices are shaped by them.

Global workspaces are 'third spaces' too. The materiality of the space is partly defined by the physical nodes of the workspace (the Exmouth Plant, the Cape Town factory) and partly by the materiality of the texts themselves. Increasingly, texts are the contexts in which people conduct their working lives. People's attitudes to texts and textual practice (both specifically, to particular texts, and generally), and to text makers (again specifically and generally) is part of the ideational dimension of workspaces. Texts 'refer' to physical locations but they do so in different ways at different locations, and they will have different effects. Texts become part of the physical location; an object or artifact to be integrated (possibly with reluctance) into the knowledge-producing practices of the people who inhabit it.

CHAPTER THREE

What counts as knowledge in knowledge economies?

Knowledge, thus conceived, is always in time, always in action, among people, and always potentiates a world in common as, once again, known in common. This account of knowledge and telling the truth represents them . . . as dialogic sequences of action in which the co-ordinating of divergent consciousness is mediated by the world they can find in common.

(SMITH)

Without knowledge an organisation would not be able to organise itself; it would be unable to maintain itself as a functioning enterprise.

(DAVENPORT AND PRUSAK)

Knowing as an active, lived experience is in a constant state of tension with knowledge as a commodity within firms and markets.

(SCARBROUGH)

INTRODUCTION

My aim in this chapter is to examine what counts as knowledge in workplaces, and in workspaces, how it comes to count, and what the implications might be for people, for organizations and for work-related education. We need to understand the forces that have shaped ideas about knowledge and skill in local workplaces if we are to have any useful understanding about how knowledge works in globally distributed workspaces, and what is involved in learning how to produce and disseminate it. In this chapter I start with a brief overview of the different ways that people think about knowledge and move to a discussion of how 'what counts' as knowledge is socially produced in local workplaces. Then I focus on what this means when something called 'knowledge' is made and distributed across global webs of production. My argument is that globally distributed workspaces demand an intensification of the textualization of knowledge that began with Taylorism, and that this shift towards the textualization of working knowledge is compounded by the digitization of knowledge through ICTs. This does not, however, mean that old forms of working knowledge are put aside and textualized knowledge replaces embodied knowledge in workplaces and workspaces. On the contrary, the knowledge embedded in systems, people and practices remains central to the success of organizations. At the same time, texts are, increasingly, the contexts in which knowledge work is done in global workspaces. A critical feature of the knowledge work of workers and workplace educators is the development (or improvisation, or innovation) of new textual practices to leverage other forms of knowledge at local sites.

CONCEPTUALIZING WORKING KNOWLEDGE

It is not always easy to get a fix on what we mean by knowledge when we talk about it in the context of work, workers, workplaces, organizations and knowledge economies. Within organizational theory alone there are many ways of understanding knowledge and no single, commonly agreed-on definition. Like many theorists, Davenport and Prusak distinguish between data, "a set of discrete, objective facts about events . . . structured records of transactions", information, "a *message*, usually in a document or an audible or visible communication" (1998: 2–3) and knowledge, which is so much more difficult to capture in a few words:

> a fluid mix of framed experience, values, contextual information, and expert insight that provides a framework for evaluating and incorporating new experiences and information . . . (: 5).

I don't want to attempt to define knowledge here, but rather, I want to use Blackler's (1995) account of five 'images of knowledge' that are used in organizational theory to give an indication of the scope of activity people are referring to when they talk about working knowledge. I should make it clear that Blackler is not suggesting that these categories represent discrete forms of knowledge; they are 'images', or ways of thinking about knowledge, found in the organizational theory literature.

'Embrained knowledge' (sometimes referred to as 'knowledge about' or 'knowledge that' and often associated with school knowledge) focuses on the abstract, relying on conceptual skill and cognitive abilities. 'Embodied knowledge' (sometimes referred to as 'knowledge how') relies heavily on immediate context and situation. This perspective on knowledge is not confined to knowledge about how to make the body 'do' something, it is centrally concerned with understanding the systems of which one is a part. Hirschorn (1984), for instance, identifies the often tacit understanding of machine systems that machine operators have as their most important knowledge. 'Encultured knowledge' is about achieving collective understandings when operating from different points of view, different industrial, professional and organizational contexts etc. 'Embedded knowledge' lies in systemic routines. It is distributed, is highly contextual, and often invisible. 'Encoded knowledge' is explicit, textual and 'generic': the knowledge of instruction manuals, policies and codes of practice.

Most approaches to working knowledge recognize how much useful knowledge is tacit, tied up in actions, routines and systems that are distributed amongst people, technologies, and organizations. Increasingly, too, knowledge is recognized as being fundamentally about people acting and reacting in their immediate communities and material contexts. Blackler, like many other commentators (Davenport and Prusak 1998; Engestrom 1999; Hildreth and Kimble 2000; Lesser, Fontaine et al 2000; Orlikowski 2002), argues that there are limits to the usefulness of developing a theory of knowledge that is disengaged from the social. The critical thing about knowledge is that, despite the common and seductive metaphor, it cannot simply be 'transmitted'; knowledge is a complex social and cultural production. Much of this research relies heavily on cognitive psychologist Jean Lave's work on situated cognition, and on the fundamental understanding that:

> knowledge is not primarily a factual commodity or compendium of facts, nor is an expert knower an encyclopaedia. Instead, knowledge takes on the character of a process of knowing (Lave 1988:175).

Coming from the perspective of sociology rather psychology, Dorothy Smith stresses the collective nature of knowledge, its association with indisputable 'truth' claims, and the fundamental 'localness' of knowing; the way in which what is recognized as knowledge is uniquely situated:

> Knowledge is not the product of a solitary Cartesian consciousness, nor is it contained in a discursive field. Sense, meaning, truth–and falsehood–are always local achievements of people whose co-ordinated and co-ordinating abilities bring about the connectedness of statements about the word and the world they index during that time, in that place and among those who participate in the social act, whether present or not (Smith 1999126–127).

These accounts of knowledge emphasize the importance of people being able to make connections with each other and to be able to 'co-ordinate' (bring together, make sense of and utilize) their understandings and their actions. While it may be tempting to treat knowledge as if it is synonymous with 'data' or 'information', that is, as if it is static, neutral and disembodied (and therefore amenable to exclusively technological 'management'), this approach implies a fundamental simplification of what is involved in making and using knowledge. Organizations will not be able to leverage the knowledge embedded in their workforces if they treat knowledge as a commodity that can be detached from the people who know it.

All in all, it seems that a focus on *knowledge* needs to be buttressed by a focus on *knowing*. Blackler argues that *knowing* is mediated, it is situated, it is provisional, it is pragmatic and it is contested. In fact it is like all other social phenomena. If we are to understand how knowledge is made and used at work, then, we need to understand that the production of knowledge is a social practice which involves dynamic relationships between people, their communities, and their (multiple and perhaps conflicting) aims and historical, material and relational contexts.

I want to work with these ideas of knowledge and knowing to develop a better understanding of what is involved when people make knowledge, and try, at least, to make their knowledge common, in globally distributed workspaces and in the local workplaces that constitute them. I begin, perhaps perversely, not by looking at the new knowledge economies and the information technologies that enable them, but at 'old' knowledge economies and the complex social and political systems that have produced and regulated knowledge in traditional workplaces. Many organizations have been both disappointed and puzzled at the patchy success they have experienced in encouraging people to disseminate knowledge, solve problems, improvise, and generally innovate, through the use of ICTs in their globally distributed workspaces (Lee and Neff 2004). My argument is that, just as local workplaces and their histories and geographies are an integral part of electronically mediated, global workspaces, so the social and political work of 'knowledge' in old economies is an integral part of the social and political work of knowledge in 'knowledge economies'; the relationships are recursive and the continuities are as important as the discontinuities.

THE POLITICS OF WHO KNOWS WHAT

Skilled work and skilled workers

Before the advent of knowledge economies and knowledge workers, debates around knowledge at work were likely to revolve around the idea of 'skill', especially as it related to and defined skilled jobs and skilled workers. The divide between skilled and unskilled work was critical because it distinguished between workers who had something distinctive and valuable to trade (knowledge about plumbing, carpentry, health care etc) and workers who were fundamentally interchangeable (no more than a disembodied pair of functioning hands, for instance, as in 'leading hand').

Despite its importance, the concept of skill has been as difficult to define as the concept of knowledge. When people talk about skill they may be referring primarily to attributes that accrue to individuals: formal apprenticeship training, to other kinds of formal training, to qualifications, to ability or to experience. They might also be referring to attributes of the job: the complexity of the job, the discretion allowed on the job. Then again, they might be referring to characteristics of the workforce: the extent to which a specific industrial or professional organization (for plumbers or brain surgeons for instance) controls the labor supply, regulates the training (the length as well as the content) and promotes specialized language and 'tools of the trade' (Noon and Blyton 1997). Unclear as the idea of skill is, it was, and still to a significant extent remains, critical to the way that workplaces and workforces are organized, and to the status and financial rewards that individuals achieve in their workplaces.

In a classic study of knowledge and skill at work at the beginning of the technological revolution Zuboff (1988) provides an extraordinarily detailed account of what was involved in 'knowing' at work. She understands all knowledge to start with the body. Our bodies are sentient; they are conscious, and reliant on the information we receive from our senses: to the sights, smells, sounds, tastes and sensations we encounter as we go about our lives. She argues that, as babies and young children, we come to know our worlds first through the information our senses give us, and we operate competently within our worlds when we are able to interpret the vast amount of complex sensory information we are continually receiving, through the lens of the specific context in which we receive it, in ways that allow us to make judgments and solve problems. This kind of knowledge often remains implicit; it is 'written in the body', and it relies for its power on our capacity to act on it with learned instinctiveness, without being overtly conscious of what we are doing.

When we learn to ride a bicycle, for instance, we learn a complex process for attending to, integrating and evaluating sensory information for continuous correction of balance which would be far too slow if we had to consciously maintain our balance as we rode. This kind of knowledge is, once learned, extremely durable. Our bodies do not forget how to ride, how to throw a ball, how to tell that the jam will set.

It is this sense that working knowledge is 'embodied'. It is, in large part, tacit, is only ever partly articulated, and when it is articulated it rarely makes sense outside the physical context in which knowledge is made and used. Zuboff describes people working in a pulp and paper mill. One man "judged the condition of paper coming off the dry roller by the sensitivity of his hair to electricity in the atmosphere around the machine". Knowledge like this is highly localized, it is related to the unique interaction between a person's individual body, the body's reaction to subtle sensory features of a specific physical context, and the capacity of the person to pay attention to the interaction and to act on the information they receive. Embodied knowledge is learned in action with other people in the immediate context of the work. Embodied knowledge is not always as explicitly physical as this example might suggest. As a young woman, learning to be a receptionist in a context in which there was likely to be a lot of conflict, I was given a detailed job description (but no explicit training) on starting work. I learned my most useful skill, to moderate the tone and pace of my voice to 'cool down' a conversation that was becoming heated, from listening to the women I worked beside and matching my vocal tone and pace to theirs.

Working knowledge has often been referred to as 'know how', literally, the capacity to know *how* to do something 'in action'. We demonstrate this kind of knowledge when we make something work. We know it in situ, but we are not necessarily able to articulate it, or even call it up when we are not *in action* in the relevant physical context. Zuboff identifies four features of this kind of 'know how':

1. *Sentience*. Action-centered skill is based upon sentient information derived from physical cues.
2. *Action-dependence*. Action-centered skill is developed in physical performance. Although in principle it may be explicit in language it typically remains unexplicated—implicit in action.
3. *Context dependence*. Action-centered skill only has meaning within the context in which its associated physical activities can occur.
4. *Personalism*. It is the individual body that takes in the situation and an individual's actions that display the required competence. There is a

felt linkage between the knower and the known. The implicit quality of the knowledge provides it with a sense of interiority, much like physical experience (: 61).

Zuboff points out that this kind of knowledge is not fragile and transitory but 'intimate, robust, detailed and implicit'. It has to be. Because bodies are frequently at risk in traditional working environments people need to have confidence in a 'knowing connection' between their bodies and the contexts in which they act.

Knowledge like this is learned, in action, in workplaces. Learning and skill are inextricably associated with 'doing the time'. It is "through the body's exertions that learning occurred, and for those who were to become skilled workers, long years of physically demanding experience was an unavoidable requirement" (: 36). This kind of learning is, as Zuboff points out, intimately connected to the *toil* of work and to the physical risk that so much work practice typically involved. In heavy industrial contexts at least, learning at work was (and perhaps still is) inextricably tied up with effort, pain and danger. High status professions like medicine also associate knowledge with work, pain and toil. Trainee doctors are still required to work very long shifts (sometimes longer that twenty-four hours at a time) in hospitals and horror stories of those shifts become part of the story of 'being a doctor'. A person's status as a skilled worker in this context is, therefore, partly associated with what they know and partly associated with what they have been through to earn their badge of office.

This way of knowing is difficult to reify in a set of discrete, or even interconnecting, 'skills', and yet the definition of skill, while a 'definitional minefield' (Noon and Blyton 1997), really matters. Skill is:

> fundamental to the status people attach to different occupations, and is frequently linked to the level of economic reward. Moreover, skill is a key factor in the structure of employment, most notably in the way it acts to reinforce the gender division of labour in society (: 78).

If specific occupational groups are to be recognized as 'skilled' in the broader community (and therefore to attract the financial and status rewards that 'skilled work' attracts) they must promote their knowledge as distinctive, scarce, and generally unintelligible to the uninitiated.

THE SOCIAL CONSTRUCTION OF KNOWLEDGE AND SKILL

It is possible, of course, to acknowledge the profoundly social nature of knowing without acknowledging that occupational groups strive to promote particular constructions of knowledge and skill in their own interests. It is equally possible to ignore the way that whole social groups contrive to construct concepts of knowledge and skill which serve their own interests and reinforce social disadvantage, both locally and globally. People, groups and cultures routinely naturalize the social construction of disadvantage, making it seem that there is a general consensus about the rank order (from unskilled to high skilled) into which skills, and therefore people, are arranged. A great deal rides on this hierarchy of skill being taken for granted. Wage and salary negotiations have traditionally relied on the understanding that, for instance, doctors are (and should be) paid more than nurses, executives more than their receptionists, warpers more than menders. And for a long while it was taken for granted that doctors, executives and warpers would have been likely to be men while nurses, receptionists and menders were overwhelmingly women. These patterns of gender differentiation are, apparently, being replicated in high-tech areas of employment growth. Production workers in Silicon Valley are mostly low-paid immigrant female workers; creative, comparatively highly paid ICT workers, like computer-game designers, are mostly men.

When debates about access and equity are taken up in popular debate the question has generally been about access, about how to ensure that women and members of minority groups have access to high-skill occupations (and the salaries and status they offer), rather than about how the judgment about what counts as skilled work have come about. Knowledge and skill are, however, fluid concepts, concepts that are routinely recruited to do the social and political work of occupational exclusion and inclusion and gendered and racialized identity construction. Knowledge and skill take shape, and assume value, at particular times and in particular places at least in part because they suit the broader social and political context, they reflect prevailing mores. As Blackler points out:

> any theory of knowing as a cultural activity must acknowledge the dynamics of domination and subordination that are a feature of everyday life (Blacker 1995: 14).

The social construction of skill is generally so normalized that it can be very difficult to detect. I want to use an example from the early 1900s, when different assumptions about categories of people prevailed, as an illustration of this point. The example is in many ways appalling, but also instructive of the way that 'what counts' as knowledge and skill can be taken for granted and blatantly used to promote the interests of one group over another. The racialization of skill is compellingly demonstrated in Mann's (2002) investigation of the piece of machinery known as the 'Nigger-Killer'. The machine, a skidder trademarked the Titanic, was used in the logging industry in Weed, northern

California, in the 1920s. Mann describes a complex political context in which transparent coercive racism coupled with attempts by white loggers to redefine themselves as a skilled workforce.

White loggers in northern California successfully differentiated themselves from African American loggers through the 'objective fact' of skill. There was nothing subtle about this, they took industrial action to assert their position and their demands to reserve the highest paid, skilled work for themselves:

> in opposition to the employment of skilled black sawyers in the best paid and highest-status job in the mill, a sit-down strike by white sawyers relegated the most experienced black workers to unskilled, low-wage, unpleasant jobs (: 480).

There were many skilled African American workers and sometimes they used their skills at work but, when they did, they were generally referred to (and paid) as 'helper' rather than as (skilled) carpenter, sawyer etc. Skill was unambiguously understood to be an attribute of race:

> Very backward races are unable to keep at any kind of work for a long time; and even the simplest forms of what we regard as unskilled work is skilled relative to them; for they have not the requisite assiduity, and they can acquire it only by a long course of training (Alfred Marshall quoted in Mann 2002: 483).

One of the ways in which white loggers asserted their skilled status was in their assumed capacity to deal with the demands of the new technologies which were being introduced to the logging industry. New technologies were expensive; they required knowledge, experience, the capacity to improvise and problem solve, and to develop new work practices. In normal circumstances, the highest-paid and most highly skilled white workers were assigned to operate the new technologies.

When the work was really dangerous, however, this rule was suspended.

> The Titanic cost huge sums of money. Operating it required physical agility, sound judgement, and quick decision making. African Americans supposedly had none of these characteristics. Moreover, many claimed that they were unsuited to the work in logging operations, only in the mills and yards . . . they were institutionally and biologically unskilled, and their wages reflected it. But the Titanic was extraordinarily dangerous, too dangerous for white labor. And so it became the 'Nigger-Killer', cheap and deadly to operate (: 488).

While, in general, white loggers argued that they were a skilled workforce because of their demonstrated abilities to deal with the technologizing workplace, this particular piece of technology was bracketed off because the physical risks to white workers were too great. Despite manifest evidence to the contrary, the loggers argued successfully that the Titanic did not require a skilled operator. 'Unskilled', and therefore cheap, African-American labor would do. Logging company owners and contactors had little reason to dispute their argument; it was in their interests also to contract the cheapest, and most compliant, labor they could get away with.

As Mann argues:

> Producing a consistently 'unskilled' workforce defined by irrelevant characteristics such as skin color is not a simple task . . . skill must be understood in such a way that the labor process, accumulation and the relations of production can change and expand without challenging the notion that these characteristics matter . . . (: 480).

The story of the 'Nigger Killer' demonstrates just how 'provisional', and how subject to prevailing social interests, the construction of knowledge and skill can be. The racialized definitions of skilled worker and skilled work were logically impossible to sustain, they relied absolutely on commonly accepted and unexamined understandings about racial characteristics. On these understandings was built a racially stratified workforce that masqueraded as a workforce stratified by skill.

The gendered construction of knowledge and skill is a feature of working life. The famous argument that:

> it is the sex of those who do the work, rather than its content, which leads to its identification as skilled or unskilled . . . skilled work is the work women don't do (Phillips and Taylor 1986)

reflects an aspect of the way knowledge and skill can be defined to reinforce and legitimate prevailing social disadvantage. It is illustrated in Cynthia Cockburn's study of the printing industry in the early 1980s, where a male compositor is quoted as saying that his work is 'men's work':

> If I said to my mates I was working with a woman, they would feel, say, oh, he's doing a woman's job—because they can see that a woman *can* do it (1983: 180).

While the pragmatic effect of women working as compositors would be likely to be a reduction in real wages for the occupational group as a whole, the implication here is that it calls into question the man's gender identity as well.

Some aspects of work, like 'emotion work' (Noon and Blyton 1997), are historically associated much more definitively with women than with men. Emotion work is work that involves the management of the emotions of the client and the worker. While occupations like call-center operator, sales assistant, waiting staff etc all involve occupational specific knowledge and skill, the distinguishing feature of the work is the demand it makes on workers to present themselves as cheerful, empathetic, and unflappable. At a time in which technological skill is valorized, it is difficult to isolate these skills as discrete, measurable, and scarce; it is not easy to see them as comprising the tool kit of the professional, or skilled, worker. In most contexts these behaviors are viewed as personal qualities rather than professional or vocational skills and, while they are crucial to the success of the worker on the job, and increasingly explicitly required in all service work, they are not rewarded financially. They are viewed as basic personal attributes that every worker in the service sector can be expected to have. Service work is the fastest growing sector of the global economy, it is generally understood to be low skilled and therefore low paid, and it is overwhelmingly performed by women.

My aim in this section has been to give an idea of the work that publicly legitimated knowledge and skill does in local workplaces and in society more generally. In local workplaces it obviously determines pay scales and wage relativities. More than that, although less visibly, it shapes, and often determines, who can do what kinds of job, whose knowledge counts and who has authority to say so, and who has authority over whom and why. Working relationships and working identities have been produced by the assumptions that underpin the regulation of occupational boundaries and hierarchies of skill, and the systems of reward that go with them.

There is a politics of who knows what, and the ways it plays out in workspaces will be significantly influenced by the ways it plays out in all the local workplaces that make up the nodes of global workspaces. Organizations certainly need knowledge, and they need knowledgeable workers, and it is dangerous to treat knowledge as a neutral, objective fact, and to ignore the contexts in which 'what counts as knowledge' is negotiated. Equally, knowledge economies do not do away with, or neutralize, local battles over what counts as knowledge and who can say so, although they may, unwisely, ignore them. Global workspaces provide a new arena in which local battles can be waged, but they are overlaid with, and complicated by, the struggles that emerge in the global workspaces themselves. What counts as 'knowledge' in cyber-place is no more neutral, and no less socially constructed, than what counts as knowledge at the Exmouth Plant or the Harbor Mill, although it may suit particular groups to suggest that it is.

MAKING KNOWLEDGE COMMON IN GLOBALLY DISTRIBUTED WORKSPACES

Knowledge in knowledge economies

From the perspective of economists at the World Bank and elsewhere, knowledge, specifically technical and design knowledge, is critical because it is the resource that leverages other resources, like cheap and flexible labor, water, land and minerals, or industrial plants. Countries need a literate, technologically capable workforce if they are to participate effectively in global economic activity. It is 'common knowledge' that:

> For countries in the vanguard of the world economy, the balance between knowledge and resources has shifted so far towards the former that knowledge has become perhaps the most important factor determining the standard of living - more than land, than tools, than labour. Today's most technologically advanced economies are truly knowledge-based (World Development Report 1999 on-line).

Claims like this one are now so common that they are rarely questioned, and yet it is difficult to imagine an historical time, or a geographical space, in which knowledge has *not* been fundamental to economic success. As Castells (1996) and many others argue, all economies have been knowledge economies in one way or another. In any Knowledge Economy, the competitiveness of nations, states, organizations, and companies depends on knowledge and information and the capacity to communicate. The Roman Empire relied on technical and design knowledge to create the transport networks that sustained the cities that were the dynamic hub of the political and economic empire, and it relied on bureaucratic, literate, communication networks to regulate and control it. The Industrial Revolution relied on fundamental technical innovations like steam power and on the technologically advanced machines that steam power made possible. What could it mean to say that, in some unique sense, ours is a Knowledge Economy? What is distinctive about the contemporary Knowledge Economy is not its reliance on knowledge, data and information, it is the extent to which, and the speed at which, knowledge is constituted and reconstituted in specific global economic networks. These processes of transformation happen partly because ICTs have a vast capacity to transform and amplify data and information and partly because global networks require that knowledge be repeatedly produced, interpreted and reinterpreted as it moves through complex, intersecting material and social contexts. Knowledge simply won't hold still.

From an ICT perspective, information and communication systems technologies change what happens to data and information once it enters a technological system:

> knowledge and information can be introduced in a technological system to create a positive feedback loop of knowledge and information, transforming it into a virtuous circle that expands by itself (Castells 2000: 3)

Once technological systems have their raw material and are set in motion they can generate more data, link up different forms of data, make more and unexpected connections and associations, and produce more information almost instantaneously and with very little human labor. Data and information expand exponentially and are constantly updated.

From a social and economic perspective, however, knowledge is important, not only because it can be commodified but, perhaps more critically, because it allows access to and participation in globally organized economic activity. I don't want to suggest that there is a single global economy, and every person, company and nation either belongs to it or is excluded from it. That is obviously not the case. It is, however, the case that participation in global economic activity is not something that we can choose to take or leave. It is true that relatively few of us are 'global workers', flying between the great global cities to engage in technologically sophisticated, highly remunerative, knowledge work. Most workers around the world are employed in local jobs, in factories and offices, schools and hospitals, and retail outlets, and engage with local people in local activities. But, this does not mean that we are unaffected by global activity. As Castells (2000) argues, ICTs have created the capacity to organize and coordinate activities on a global scale, to create multiple, intersecting networks. These networks operate across national, corporate, industry and sectoral boundaries. All jobs, even the most local, are affected by these frameworks in varying degrees: 'we all depend on some key activities which are organized globally' (Castells 2000: 4). The economy is a network economy. It is not based on individual corporations, states or governments, but on networks:

> [the economy] is organized on networks of firms, segments of governments and segments of the public sector. . . . The operating unit is the network that is organized in a particular business project to implement the particular strategic decision (: 4).

To understand what knowledge is and does in knowledge economies we need to focus on the network. The globally distributed supply chain that produces the global car, described in Chapter Two, is just such a network, and it intersects with other networks organized around the production and marketing of various other world products and services. Mary and Bill make automotive textiles at the Exmouth Plant. Theirs are local jobs, and they mostly work with local people and operate under local labor regulations in local contexts. Their working lives are, nonetheless, intimately affected by the networking demands of that global supply chain. As the chapters which follow show, they are global workers in the sense that their local practice must adapt to accommodate the social and work practices that integrate them into global networks.

To say that successful contemporary economies are founded on knowledge is, then, to say that in order to participate in economic activity (manufacturing, service, technological . . .) a state or a company or an individual must have the knowledge to insert itself into, create and sustain transient, globally organized technologically enabled networks. Without this capacity no amount of land, labor, raw materials, or local knowledge and skill, can be mobilized as a resource. From this perspective, the critical knowledge, the knowledge that leverages other knowledge and skill, is the knowledge that enables us to make and sustain connections in global webs of production. The knowledge that is fundamental to effective participation in global economic activity is the knowledge that is required to connect, transform and translate knowledge across contexts; the knowledge that allows us to create and maintain global work*spaces* without immobilizing local work*places*.

Like economies, organizations, too, have always relied on knowledge. As Davenport and Prusak point out, all successful companies are and have been 'Knowledge Organizations':

> Without knowledge an organization would not be able to organise itself; it would be unable to maintain itself as a functioning enterprise (Davenport and Prusak 1998:52).

Knowledge has always driven economies and organizations. What is distinctive about contemporary corporations is that they must make and trade their knowledge in global networks in order to participate in global economic activity. This means relaxing their hold on corporate knowledge and working cooperatively with other participants in a high pressure 'just-in-time' environment. To operate effectively corporations must be able to insert bits of themselves (individuals, departments, project teams etc) into constantly reorganizing, globally structured, networks (made up of parts of other companies, government departments, Non-Government Organizations etc) if they are to operate in global networks of production. More or less permanent internal structures, like departments, become less important, and transient cross-departmental or external structures, like cross-functional teams (the Action Learning Teams that Bill and Mary belong to, for instance) and project teams, become more important. Internal corporate organization and governance competes with external organization and governance (like ISO) to regulate local work practice and working relationships.

Segments of companies, and formal and informal alliances of workers within and between companies, must be able to operate relatively autonomously to solve problems and innovate quickly. At the same time, if their knowledge-building is to be effective, they will need to refer simultaneously inwards, to the practices and processes of their company, and outwards to take account of the practices and processes of other participants in the network, and the intersecting organizational constraints. This is quite a balancing act, and it requires people operating at the borders to develop sophisticated new ways of communicating that involve interpreting local knowledge to a global audience and interpreting the global knowledge back to the local community. Increasingly, embodied, embedded, encultured knowledge needs to be turned into texts (words, diagrams, symbols etc) in order to move it around the network.

THE TEXTUALIZATION OF KNOWLEDGE

The textualization of working knowledge is not an entirely new phenomenon. It began with Taylorism as a means of controlling the processes of work (Braverman 1974):

> The managers assume . . . the burden of gathering together all the traditional knowledge which in the past has been possessed by the workmen (sic) and then classifying, tabulating, and reducing this knowledge to rules, laws and formulae (: 112).

The idea was that if the processes of work could be comprehensively described then any person could do the job by reading the manual and following the standard operating procedures without deviating. Any discretion workers may have had in doing their jobs was compromised, but, perhaps more significantly, any bargaining power they might have, individually or collectively, was effectively discounted because they could, apparently, be replaced by any other physically able person who could read the instructions. While the textualization of work has been around for a long time, it has accelerated dramatically since the advent of digitization.

Digitization means that people around the world (not everywhere, but in many key places) can have access to the same data and information within moments, and can receive updates regularly, but this in itself does not make global networks possible, and it certainly does not make them effective. Knowledge production and diffusion in globally distributed networks rely on people being able to *connect* with each other, being able to interpret for each other, evaluate knowledge claims and engage local communities in knowledge-building activities. These connections coordinate knowledge-making activity, they allow knowledge to 'work'. The productive coordination of knowledge-making activity rests on textual practice, on the reading and writing involved in E-mail, text messaging, computer-mediated work environments (like those described in Chapter Five) and hard-copy print, and the talking and listening involved in telephone, video conferencing etc. Text is the context in which work is done and work*spaces* are in fundamental ways textual spaces. Since work*places* are integral parts of workspaces, they are becoming textual places too.

The textualization of knowledge at work is not a simple matter. It involves decisions which reformulate knowledge, which foreground different aspects of knowledge and which inevitably reify some forms of knowledge as universal truths and others as not. Translation between platforms, for instance, digitizing a written text like a memo, or report, involves deciding what 'counts' as the information of the document. Brown and Duguid (2000) tell the story of an archivist sniffing letters written during the plague years to determine which villages the plague had reached. Although the letters were commercial, and rarely mentioned the plague directly, the smell of vinegar (which lingered after 200 years) indicated that the plague had reached the writer's village and the letter had been disinfected. The most well-intentioned and meticulous digitization of these letters would lose that information if the digitizer did not know to sniff for vinegar.

Within global workspaces, different communication technologies are complementary rather than competitive. Print texts, which continue to play a pivotal role in workspaces, are stable and immutable, while other forms of digital communication, like web pages, are more fluid. As knowledge shifts between different communicative forms it changes. Brown and Duguid argue that, within workspaces, "[d]ocuments do not merely carry information, they help make it, validate it and structure it" (: 190).

Given that all knowledge can be understood as social, and what counts as knowledge is already socially produced and reflects and reinforces social worlds, and given that global workspaces are no less social for being primarily digital, a lot is at stake when knowledge is shifted from mode to mode and context to context. These acts of translation are not neutral, they involve prioritizing some knowledge over others, and transforming some forms of knowledge into other forms so that they are visible, and so that they have legitimacy, in a range of different contexts. Acts of translation do not happen without a fight. As the following chapters show, what 'counts' as knowledge is fiercely contested, in the act of translation, at local workplaces and in global workspaces.

What counts as knowledge, and who has control over it, is also contested in organizations. Knowledge Management in organizations is no longer the exclusive province of IT departments; Brown and Duguid (2000) report 'sounds of fighting between the Information Technology and the Human Resource departments over who "owns" knowledge management' (: 118). Progressively, debates about Knowledge and Knowledge Management in organizations have shifted from a focus on information and communication technologies and the problem of capturing, codifying and transferring data and information, to a focus on the social conditions, the social knowledge and the social skills, that promote the production and use of working knowledge in globally distributed networks. Within this context ICTs are viewed, not as neutral data-processing technologies, but as an integral part of the social conditions of knowledge production at local sites and as critical in joining up local workspaces with other sites, and with the cities, regions, nations and global networks in which they are embedded.

This is particularly complex because the knowledge is disembedded when it is textualized and needs to be reanimated in what is generally an entirely textual environment. In one sense we can see that the textualization of knowledge means that it is "exteriorised to the knower" (Jean-François Lyotard 1984) in much the same way as it was with Taylorism, an issue that Bill and the team struggle with in Chapter Six. In another sense, however:

> [t]he importance of people as creators and carriers of knowledge is forcing organizations to realize that knowledge lies less in its databases than in its people (Brown and Duguid: 121).

Unlike information, knowledge is hard to detach from people; it is as Zuboff pointed out, deeply embedded in practice and often cannot be 'actioned' outside the context in which it is made. The transfer of 'best practice' has proved notoriously difficult. As AFM found, it is difficult enough to transfer practice from the Harbor Mill to the Exmouth Plant, let alone across corporate or national boundaries.

This shift to the idea of knowledge as tied up with people and their practices, in context, is fraught. It involves corporations in de-emphasizing knowledge as explicit, measurable data and information, amenable to the application of various ICTs, and re-emphasizing knowledge as tacit, situated, distributed, and embedded in social contexts which include ICTs. While the focus is shifting uneasily from 'knowledge' to 'knowing', the tension between knowledge and knowing is often a problem for global corporations. On the one hand, they rely absolutely on knowledge; the production of knowledge (in the sense of data certainly, but more importantly in the sense of innovation, new product design, problem solving and the like) is core business in any global company. The communication of knowledge across globally dispersed workspaces is fundamental to the healthy functioning of an organization and the commodification of knowledge is critical to an organization's economic viability. On the other hand, the social practices that nurture knowledge production and communication in globally dispersed companies are not the same social practices that sustain managerial control or support the exploitation of knowledge as an economic product traded in a global market. In fact, the practices that cultivate the production of knowledge often seem to threaten organizational chaos and anarchy, or at least promote personal and group autonomy at the expense of corporate goals and

> *knowing* as an active, lived experience is in a constant state of tension with *knowledge* as a commodity within firms and markets (Scarbrough 1999:6).

Organizations have needed to find organizational structures that sustain and promote the kinds of collaboration that build new knowledge. The idea of 'Communities of Practice' (building on the work of Lave and Wenger 1991) has played a pivotal role in this process. Communities of practice organize themselves around a joint enterprise, they are bound together as a social entity, and they produce a 'shared repertoire' of resources over time. The spatial metaphor 'common ground' is used to indicate the shared resources and shared interests of the group (Hildreth and Kimble 2004). The concept of individual communities of practice is giving way to the more extensive concept of 'Knowledge Networks', a concept which acknowledges that membership of communities of practice overlap and that work is temporally and geographically (and, with outsourcing, organizationally) distributed.

Communities of practice rely on shared culture to produce and disseminate knowledge and, as Gee explains, the culture of a community of practice is both its strength and it weakness, from a corporate point of view:

> Culture, if it 'takes', has the primary problem that it ultimately bonds people to each other through unconscious allegiance to shared sense-making practices (often stories) that are all about being a certain kind of person . . . cultures create bonds between people that often disallow the flexible regroupings, boundary crossings, project turnovers, and identity shape-shifting that the new capitalism demands. They also create the danger that we see each other as the same 'kind' of person, our loyalties and allegiances will be to each other, and not primarily to the firm, especially as it challenges our trust (eg by laying some of us off) (Gee 1997:77).

Global corporations support the idea of *certain kinds* of communities of practice because it gets them out of the seemingly intractable problem of reconciling the need for community with the need for control. Gee (1997) argues that the kinds of communities of practice supported by global corporations place the emphasis on *practice* rather than on community in its traditional sense of shared values and beliefs and identities. Individuals are invited, or instructed, to associate themselves primarily with the practice, and only secondarily with the other individuals, the community:

> The solution in the new capitalism is to bond people, cognitively and affectively, not first and foremost to each other, but to what I will call a *practice*. In the process a community of practice is created (Gee 1997:77).

In the contemporary global corporate world, practice persists, although individuals may come and go. Gee argues that global corporations attempt to exert control over their workforces by foregrounding practice as a means of joining people together and generating new knowledge. At the same time, the traditional collegial and collective relationships of trust are discouraged and undermined.

The power of ICTs in this process is important:

> As well as facilitating CoPs [Communities of Practice], IT plays an important role in developing measures and metrics for supporting CoPs as a value adding business resource. *The ability to track CoP activity provides an additional value-adding role for IT* (Lee and Neff 2004 my emphasis).

Information Technology allows organizations to track, monitor and archive every digital utterance and exchange a community of practice might generate in the course of a project. It promises a level of surveillance that Taylor could only dream of.

THE COLLABORATIVE CONSTRUCTION OF A KNOWLEDGE WORKER

Most of the discussion around the knowledge problem has focussed on 'knowledge workers', that group of workers (Scarbrough argues it constitutes about 29% of the workforce) recognized by organizations as primarily concerned with the production and application of knowledge in their work. Within these analyses, the major part of the workforce is assumed to be undertaking routine work within traditional management structures. Elite knowledge workers, on the other hand, are generally understood to challenge more conventional management practices, partly because of the high degree of autonomy they demand (and often receive) individually and as groups, and partly because of the importance of the close and durable bonds they have with colleagues inside and outside the company.

Increasingly, however, the cut-and-dried distinction between 'knowledge workers' and other workers is difficult to sustain. In global networks of interaction companies need workers at every level to generate new knowledge which, at the very least, integrates global corporate knowledge about how things are done with local knowledge and local practice. It can be pragmatic for some organizations to promote some members of their workforce as 'knowledge workers', and strategic, not to mention flattering, for some individuals to accept that designation, but the identities associated with the knowledge worker are often collaboratively constructed.

Mukerji's (1998) study of the construction of knowledge in a science laboratory demonstrates how locally generated knowledge is only legitimated when it is produced and interpreted in the light of global discourses and through the collaborative construction of a single 'scientific knower'. More than most kinds of labour, labour in a scientific laboratory seems to be unproblematically associated with the production of knowledge. There are now many studies of the sociology of scientific knowledge, studies which demonstrate that such knowledge develops collectively in the relationships and interactions of members of the scientific team. Mukerji is not concerned so much with the way knowledge is constructed in the teams as the way the knowledge the team creates is presented to the outside world. She is interested in the way that scientific teams make sense of their findings and invest them with authority—the way they ensure that they become legitimated as 'scientific knowledge' and are endowed with all the prestige and authority Western society currently bestows on science. Her task is to uncover what she calls the cultural assumptions that scientists use, within which certain sets of experiences and practices come to be legitimated as scientific knowledge, while other experiences and practices do not. She argues that scientists reproduce the idea of the *scientific genius* as a means of investing their work with an aura of truth. In this way she links local processes of cognition to the cultural production of 'knowledge':

> Part of what research teams do is collectively produce a 'scientific knower' at the same time as they produce scientific knowledge. They give authority to the knowledge they collectively produce by making it the brain-child of a 'great man' (almost always gendered as male)–someone who can think better than ordinary people because he is in fact expressing the thoughts of a group. They organise their group life to construct an authorship that makes the results of their work socially more authoritative; they give their collective actions a singular identity, centred around the chief-scientist-knower (: 257–258).

In other words, this study demonstrates how a particular group of scientists call on global discourses of science at their local site in order to ensure that their work is acknowledged as, legitimate, 'good science', in their global scientific community.

Scientists are not alone in legitimating their work by associating it with an 'expert knower' who 'thinks better' than ordinary people. Blackler (1995) points out that the idea of professional knowledge has traditionally allowed professionals, like doctors and lawyers, to 'black box' their skills, and Mats Alvesson (1993) points out the concepts of knowledge work and knowledge worker provide the same apparent authority and power for a new generation of workers. The social effect of these terms legitimises particular divisions of labour and particular systems of reward in much the same way as it always has done. It is likely that most 'knowledge workers' are backed up by a team of 'knowledgeable workers', like Mary and Bill and the others, who contribute to the construction of what counts as legitimate knowledge and who get to be an 'expert knower'.

CONCLUSION

What I've tried to do in this chapter is to begin to identify some of the forces that have shaped ideas about knowledge and skill in local workplaces so we can begin to understand how knowledge works in globally distributed workspaces and what is involved in learning how to produce and disseminate it. I've argued that, when knowledge is textualized, as it is when it is translated from one platform to another and one site to another, it not transmitted but, rather, reconstituted, and what it is and what it signifies becomes something different. All of this is a fairly abstract argument. In the chapter that follow I try to pin down the microprocesses of knowledge production in workspaces, and the role that textual practice plays.

CHAPTER FOUR

Solving problems by the book

> Readers are travellers; they move across lands belonging to someone else. . . . Writing accumulates, stocks up, resists time by the establishment of a place and multiplies its production through the expansion of reproduction. Reading takes no measure against the erosion of time (one forgets oneself and also forgets), it does not keep what it acquires, or it does so poorly, and each of the places through which it passes is a repetition of a lost paradise.
>
> (DE CERTEAU)

> Paper documents have proved more resistant than many of their antagonists (or defenders) expected. Paper is not, moreover, simply hanging on. New avenues for paper documents continue to develop, while its resourcefulness and complementary properties, though previously dismissed, are now becoming an asset for digital technologies.
>
> (BROWN AND DUGUID)

INTRODUCTION

In this chapter, and in the chapter that follows, I want to look at the ways that ICT-enabled globalization is reshaping traditional textual and knowledge-building work practices, like problem solving, and the ways in which traditional work practices are shaping new and emerging work practices. The focus of a good deal of education literature has been on the ways that technologies are profoundly reshaping the world, rendering physical space and established identities irrelevant. In space, it is said, distance dies, and people can become whoever they want to be. Virtual worlds, it is argued, demand new textual practices, new literacies. Research focuses on comparing on-line and off-line communities and activities, and generates calls for schools and other educational institutions to teach the textual practices of virtual worlds.

My focus in this book is on the textual practice of knowledge production in contemporary workplaces, and that focus complicates the distinction between on-line and off-line and (as I have already argued) the associated distinction between local and global. If we are looking at knowledge production at work it is simply not helpful to distinguish between on-line communication and any other kind. My argument here is that on-line communication is part and parcel of all communication in workplaces, and face-to-face and print-based communication is part and parcel of ICTs. Talk, print and technologies are integrated into workplaces, and they are mutually dependent. Print texts occupy a powerful place in contemporary local workplaces, at least as powerful, I would argue, as the place they have occupied over the past two hundred years. The stubborn persistence of the fax (flawed technology according to most ICT experts) and of hard-copy printouts of electronic documents, attests to the stubborn durability of print and its capacity as a medium to transform, accommodate and adapt. In fact, it is print that makes a good deal of electronic data usable in technologically hybrid workspaces. To understand how technology operates in workplaces we need to understand how print operates and how people operate with various textual forms and communication media. As Graham argues:

> New information technologies, in short, actually resonate with, and are bound up in, the active construction of space and place, rather than making it redundant (1998: 174).

Despite my commitment to an analysis of textual practice in workplaces that integrates all forms of textual practice as constitutive of knowledge production, I am focusing this chapter on print-based texts and the following chapter on virtual environments. Given this arbitrary division, it is especially important that I emphasize the interconnections between the two. The print-based texts that I talk about in this chapter have been called into being by the imperatives of global webs of production; they are updated and adapted, and distributed simultaneously around the world, electronically. As I demonstrate in this chapter, print-based texts also have a profound effect on the face-to-face textual practices of local workplaces; so talk, print and electronic texts are all reliant on each other and shaped by each other. In the same way, the 'virtual workspaces' that I talk about in the next chapter are inserted into existing workplaces and rely heavily on traditional textual resources, face-to-face and telephone talk, print-based manuals and 'back of the envelope' diagrams. While they are often presented as seamless, context-independent, problem-solving textual environments, their knowledge-building capacities rely on people being able to mobilize their other, local, resources to improvise the textual practices that join up the gaps and make the connections that allow virtual workplaces to work.

This chapter focuses on the ways in which print texts shape knowledge production in local workplaces, how it happens, and what happens when they do.

TEXTS, TECHNOLOGIES AND KNOWLEDGE PRODUCTION

To reiterate, globally distributed workspaces are communities made up of people and technologies connected by practices that are, in the first instance at least, textually realised and regulated. Global workspaces are webs of textual practice, created and maintained by people reading and writing and using technologies to transform print through digitization. These processes of transformation amplify the power of texts in workplaces by extending their depth and reach and obscuring the contexts in which they were produced so that a document produced in Utah seems to apply equally to rural or urban workers in Cape Town, Exmouth or Harborside. It cannot, of course. Texts like these need to be translated so that they make some kind of sense in radically different environments. Significant print texts, emanating from remote 'Head Offices', generate new textual practices at local sites, practices (like team meetings) that are designed to create a join between individual local workplaces and work practices.

The relationships between individual local workplaces and the global workspaces of which they are a part is complex and contingent. Regulatory texts attempt to formalize and stabilize these relationships, describing in surprising detail the processes by which particular products are to be manufactured, the specifications of the raw materials to be used and the standards to be achieved. It seems that global corporations are trying to impose by text the kinds of organizational roles, relationships, responsibilities and practices they would assert if the workers were in the same physical space. They attempt to homogenize and standardize practice. In this chapter I want to suggest that, when things go well, they may make work practices compatible, and texts comprehensible, but they do not 'standardize' them, they remain stubbornly located in local conditions, histories and geographies.

I have referred already to the attention that literacy has received as a means of moving people and nations into global knowledge economies. Labor Organization, the United Nations and the World Bank, have initiated programs to 'eradicate' illiteracy in developing countries, paying attention to both universal schooling and workplace and workforce education programs. Developed countries like Australia, Canada and the USA have also instituted workplace English literacy programs, recognizing that, while many workers were 'literate' in a traditional sense, they were not reading and writing in the ways global corporations seemed to require. As Glynda Hull (2000) has pointed out, distinctions between 'workplace literacy' and 'school-based literacy' have generally distinguished between the applied nature of reading and writing at work (reading to do) and the information or knowledge-based nature of school-based literacy (reading to know). These distinctions have been reinforced by distinctions between the traditional conceptualizations of working knowledge discussed in Chapter Three and ways of conceptualizing knowledge generally called upon in research on schooled knowledge. These apparently cut-and-dried distinctions are, however, increasingly compromised as we get a clearer sense of what 'knowing' might actually involve (at work and at school), and as we see the ways that corporate organizational texts are deployed, not simply to transfer data and information, but also to change local work practice and to regulate people and practice at a distance.

The quantity and variety of print and electronic texts that now saturate many local workplaces consign an often unacknowledged burden to many workers and challenges established working identities and working relationships in many workplaces. These changes are not confined to the workplace. They spill over into broader social and community life, disrupting the way people see themselves, their expertise, and their place in their immediate world. In Sholtz and Prinsloo's (2001) account of a Cape Town hydraulics factory, for instance, they draw attention to the shifting status of Farrida, who was appointed first as a machinist then as the first woman Team Leader in the company. Farrida is a relatively new employee who has advanced rapidly through the ranks from her initial appointment as a machinist to that of Team Leader. Farrida became Team Leader because the previous appointee resigned from the position:

> That guy over there . . . he used to be a Team Leader. But then it became too much paper work so he threw his sweater . . . his Team Leader's sweater . . . in the rubbish bin. Then he didn't want to be Team Leader anymore. Then he said it was the Boers (Afrikaans farmers/White men/bosses) but he was wrong . . . and all that, but it was because he couldn't write (: 711).

Sholtz and Prinsloo point out that, contrary to this claim, the original Team Leader could indeed read and write, but not in the ways he was required to do since the South African company's incorporation into a global operation. The Team Leader role had been transformed into one that required fluency in spoken and written English. The process of her appointment was directly related to Farrida's ability and willingness to gain mastery of the new genres, interpret them to the members of her team, liaise between her team and management, and to do so in spoken and written English.

The tensions that were created as traditional, male, machine-based, expertise was supplanted by female, text-based expertise challenged traditional gender relationships but reinforced the traditional divide between management and workers and exacerbated cultural and linguistic tensions:

> If we understand reading and writing at work not simply as basic skills, but as a more varied range of skills, attitudes, and identities then we get a better handle on what these tensions are about (: 711).

The sorting, prioritizing, production and interpretation of written texts has become a significant part of the working day for many people. The quantity and representational form of written texts are important, but only part of the story and not the most important part. What matters most to people at work is the way new texts (which may in many ways look entirely conventional) are animated in local workplaces. What matters most is the way people integrate them and interpret them in the course of their working lives.

For education, and for eductors, this distinction is particularly important, but it is particularly troublesome too. If the focus is on the quantity and genre characteristics of written texts then the task of work-related education, and workplace English literacy programs especially, is to identify the salient characteristics of texts and teach people to read and reproduce them. This is reassuringly manageable in very many contexts, or at least it looks as if it is. If, on the other hand, the focus is on the ways people continue to weave new written texts into the complex social fabric of their working lives, then the education task is less clear and much more complicated. It is, however, the complex task that is the one that needs doing.

What people need to know about texts is not so much how to read them, or how to write them, as how to *use* them. Knowing how to use a text involves understanding how the written text joins up the local workplace with the global web of production in a network of practice. At least as significantly, however, it involves understanding how to use these written texts to have some control over work. It is in the animation of the important written texts of their workplaces that people, individually and even more so collectively, can claim some agency in the way their working lives are configured locally.

Hull (2000) tells the story of Mr San, a Team Leader in a circuit manufacturing factory in the USA. She reports that many workers in the factory resented the literacy tasks, the new roles, and the additional responsibilities, associated with the imposition of a 'self-directed work team' structure. They disassociated themselves from new literacy demands like 'presentations', performing them in a disengaged way and explicitly discouraging questions, debate and collaborative 'problem solving'.

Mr San, however, used the opportunity provided by a training meeting (intended to develop the new workplace literacy practices of the team) to appropriate the meeting, draw attention to a significant problem for his team, and provide a solution:

> Yet Mr San commandeered this practice session to demonstrate how he had solved a very important literacy-related work problem . . . and in doing so he showed how he had used literacy to resist authority. He had refused to supply the necessary documentation for calculating productivity and quality scores until the correct [information] had been provided (2000: 650).

Hull attributes Mr San's capacity to appropriate the new textual practices of the workplace to Mr San's established identity on the factory floor as 'a kind of father figure, a protector and an advocate' for his team. He used his established social role in the workplace as a resource. It allowed him to make sense of the new textual practices, master them, and recruit them to his own purposes. In this case at least, globally imposed textual practices are called on as resources in the politics of the local workplace:

> Mr San had found a space for participating in the work of teams, as the company desired, and for shaping team work toward the collective good (652).

In Hull's account, Mr San did not merely reproduce the required text, he appropriated it and *used* it. As people like Mr San learn how to use global workplace texts they may be able more consciously to influence the micro processes of globalization as they are played out on the factory floors, office blocks, airport lounges and coffee shops that constitute the contemporary workspace.

What matters most about the effects of written text on workplaces is the way they are incorporated into working life, changing the social and power relationships and inflecting work practice and knowledge production in obvious and in unobtrusive ways. Work-related education could have a powerful effect here, if only we knew what to do.

SOLVING PROBLEMS AND KNOWLEDGE PRODUCTION

Two hours before the report is due the computer in the home office mysteriously stops talking to the printer. The yarn required for a new fabric design is unavailable, and the specifications of the replacement are not identical; the warping machine has jammed on the fabric, and the order is already behind schedule. An engine has cut out just as the aircraft is about to begin descent. A patient with an unusual heart condition needs a particular kind of stent that hasn't yet been designed; an existing stent could be modified, it might save him, might kill him. Workplaces are full of problems that need solving, in fact problem solving is the generic form of knowledge production; it is the engine of innovation, shared by surgeons and mechanics and nurses and process workers and farmers all over the world. For many, perhaps most people, problem solving is the creative edge of their work, the work where they call on their own unique experience and knowledge and the experience and knowledge of their colleagues to produce new solutions. Businesses rely on workers' being able to identify problems and solve them quickly, ensuring, where possible, that the solution does not generate more problems than it solves. Paradoxically, problem solving is the point at which most workers have most agency. There is an urgent need to innovate—to do something differently in order to cope with a new situation or an old situation that can no longer be tolerated. Innovation, or new knowledge, is required; you can't just do things 'by the book'. Despite the counter-intuitive nature of this proposal, many companies seem to be trying to do just that, to regulate the way people think about and solve problems by designating specific problem solving processes. A written text is mandated to regulate how people deal with the unpredictable.

This chapter takes a close look at what happens when a critical written text is introduced into a specific local workplace. I pay attention to how the people and the organizations go about making sense of it and integrating it into their work practice, how some of the old conflicts play out in the process and the new ones that emerge, the reconfiguration of social relationships, communities of practice and authority relations and the old/new ways of doing the work that emerge in the process. There is nothing predictable about the process, except that it is always uniquely inflected by the local workplace.

The Lacemaker

Bill is a Lacemaker. He was apprenticed to one of the large machine-lacemaking companies in southern England when he was a young teenager in the early

1950s. Lacemaking sounds like delicate, clean, genteel work, but machine lacemaking, even as late as the fifties, was physically demanding, heavy and dirty. Apprentice Lacemakers were trained by Master Lacemakers. They were taught the principles on which the different lacemaking machines were built, how the machines would handle the different kinds of yarns, how humidity or extreme cold would affect the machine and the fabric when it was finished. Many of the men and boys in the town worked in the lacemaking factories, and they admired skill in coaxing the machines to produce infinitely intricate patterns. Traditionally, Machine Lacemakers were acknowledged as the elite of the town's workforce. Bill is proud that as an apprentice he, with other apprentices, made an exquisite panel of the lace in Princess Elizabeth's wedding dress. In the town where Bill grew up lacemaking was unambiguously men's work. Not only were the machines heavy and unwieldy, the oil and grease left workers filthy at the end of the day and the technical complexity of the machines meant that they frequently broke down and had to be cajoled back into life. Women were employed to finish the lace, but it was the men who knew the machines intimately, controlled production, and generally set the rhythms of working life in the factory.

Bill came to Australia with his family on the Ten Pound Migrant scheme in the early 1960s. He worked in textile factories and settled and raised his family in the industrial northern suburbs of Melbourne near the factories where he worked. His early training in the structure and workings of various classes of machines was highly valued by employers in Australia. Textile production was a historically important part of the Victorian manufacturing industry, but the industry was changing, and increasingly feeling competitive pressure from Asia. Skilled workers, especially those who understood the new imported machines, were hard to find. A worker who could operate the machines efficiently and fix them quickly when they broke down saved the company many hours of downtime. Getting orders out on time and on budget was crucial to the survival of the company. Bill knew the machines from his work in England, liked to understand how they worked and took pride in his skill in diagnosing problems and inventing solutions so that orders were completed 'on time' and 'on budget'. He has been working at AFM for nearly forty years, poached from a rival company soon after he arrived in Australia. He has taken several supervisory positions at AFM and is currently Warping Shed Supervisor at the Exmouth Plant. Bill reads and writes in his role as Warping Shed Supervisor, but his textual practice is heavily embedded in the material conditions and local work practice of the Warping Shed. His work station is surrounded by production schedules, his in-tray overflowing with orders, specifications, occupational health and safety bulletins and handwritten notes from colleagues. The critical textual innovation in the Warping Shed, the one which maintains connections between local operators working in shifts in a very noisy environment is the yellow 'Post-it Note'. It is affixed to machines when they break down, to yarn to indicate which job it is for, to operators' chairs to inform them of phone messages and to computer terminals to alert operators to data judged to be incorrect or out of date.

When we meet Bill he is seated around the board room table on the Fluffy Floor of the Exmouth Plant, taking part in an Action Learning Team Meeting. There are sixteen people sitting around the table, and Bill knows them all. Some are from the Exmouth Plant, others are from the Harbor Mill. So, what is an Action Learning Team, and why has AFM management decided it needs one?

The Action Learning Team

In Action Learning Teams participants identify a problem they would like to solve or an outcome they would like to achieve and learn how to do it 'in action'—while they are engaged in the process of doing it. In the process of solving the problem they have identified, the team (ideally) also learns how to function effectively as a team. The Action Learning Team process is intended to develop individual and collective team-building skills which will provide a foundation for collaborative knowledge production. In this particular case, the professional development program is funded by the Australian federal government and has been implemented by AFM because some, but by no means all, of management are concerned about AFM's performance in the increasingly competitive automotive market.

Sally chairs the Action Learning Team meeting. She and Margaret are both Enterprise Based Teachers (EBTs) employed by the local Technical and Further Education (TAFE) Institute and contracted to work at AFM. Sally, based at the Harbor Mill, trained as a secondary English teacher; Margaret, based at the Exmouth Plant, trained as a primary teacher. They moved into adult education as basic literacy teachers, but now they are both 'enterprise based'. This means that they are based at specific workplaces for a period of time, integrating their training into the idiosyncratic structures and practices of individual workplaces. It has been many years since they have taught basic English literacy to immigrant process workers. For the last several years their work has involved developing individual and group training programs for apparently literate workers in companies, like AFM, that are undergoing significant restructuring in response to the lowering of tariffs and the globalization of markets. Most of their work is concerned, implicitly or explicitly, with facilitating management restructure in companies, especially in helping establish cross-functional, knowledge-producing teams. They maintain that their work, while no longer basic literacy, is still, fundamentally, literacy work.

Sally and Margaret negotiate specific workplace education programs with company management (mostly) and employees (sometimes) within the guidelines provided by various federal government funding bodies. They are experienced EBTs, each having worked in a variety of manufacturing contexts for over ten years, and in this case are contracted (paid for partly by AFM and partly by a federal government-funded workplace education program) to establish and support an Action Learning Team and to provide other kinds of English language and literacy support to the employees at AFM.

The Action Learning Team at AFM consists of Sally, Margaret and fourteen Supervisors and Managers of various departments. The general purpose of the team is to establish and promote team-based practices at AFM, an organizational structure that neither the management nor the employees really want. The imperative for team-based organization is externally imposed. The global automotive industry requires closely integrated component suppliers to provide some protection to fragile supply chains. One of the ways this is achieved is by ensuring that all suppliers adopt certain organizational practices, particularly cross-functional teams and team-based problem solving, if they are to continue to be viewed as preferred suppliers. The idea is that, if all workers in the supply chains have learned to solve problems in the same ways, then they can be plugged into various globally distributed teams as needed with minimal disruption.

At the time that the Action Learning Team was established, Autoco, a major international automotive assembler, had notified AFM that their score on the Autoco Quality Assessment was unacceptably low and that changes to local work practices, especially in the area of team-based, structured problem solving, must be made immediately. The requirements were detailed in the *Autoco Quality Manual*, a set of rating scales against which Autoco departments and supplier companies ('trading partners') are judged. One of the rating scales is reproduced in Figure 1 below. According to Autoco, AFM is not achieving an adequate score on this particular scale–in fact it has barely rated a 2 at the last assessment and needs to rate at least a 3 if it is to retain Autoco's custom.

As a response to this situation, the management of AFM, in consultation with Sally and Margaret, agreed that the main task of the Action Learning Team should be to produce an Eight Step Problem Solving Manual for use by employees at AFM. The Manual is intended to guide employees at AFM in the establishment of Eight Step Problem Solving Teams. Adapted from a model provided by Jim, the Human Resource Manager, it describes and prescribes in detail the eight steps to be followed when a problem has been identified and needs to be solved. It is intended that successful Eight Step Plans will be generated and documented by the Action Learning Team and these will comprise the AFM Eight Step Problem Solving Manual. The existence of the Eight Step Problem Solving Manual will go some way to improving AFM's score on the *Autoco Quality Manual*. So, the team is using a generic manual to produce another local manual, and local practice is tied into a complex textual regulation network. Any confusion readers might experience in getting these manuals and plans straight in their minds mirrors the confusion expressed by members of the Action Learning Team.

'Problem solving' as knowledge production

There is nothing special about the *Autoco Quality Manual*, documents just like it are produced every day by most global companies but especially in the automotive industry, and, although they are updated regularly, their basic purpose doesn't change. AFM deals routinely with three major car assembly companies, and each of them mandates certain organizational practices that reach deep into the structure of local companies. AFM is not alone in that it must comply with the slightly different mandated practices of three manuals, from the three different assemblers it supplies, simultaneously. The *Autoco Quality Manual* was produced at the Engineering Department of the Head Office of Autoco in the USA. It is intended to apply to all the departments of Autoco in the United States and elsewhere, and all the supplier companies in Autoco's various supply chains. Chapter Two identified one of the problems faced by companies producing the global car: in outsourcing the production of critical components, the company risks ceding control of the final product to loosely affiliated remote organizations. The *Autoco Quality Manual* is an attempt to assert and retain control by controlling the processes and relationships of the supplier companies. It tries to do this partly by reifying certain locally situated practices (like the single best way to solve a problem) as 'best practice'. Companies are regularly assessed by Autoco, and at its last assessment, AFM was failing on its problem-solving rating (the scale is reproduced below), it was not using 'structured problem solving' (at least as it is described in the Manual) at all.

RATING PROCEDURE

4. Is there effective use of structured problem solving (eg. 8D)? Are root causes determined and verified?

RATING

0. No problem solving	There is no evidence of structured problem solving
1. Only for the customer	Structured problem solving is used, but only at the request of the customer
2. For internal issues	Structured problem solving is used for internal problems such as those identified in [the Manual] as well as for external problems
3. Use of teams and tools	Structured problem solving employs team participation and all of the basic problem solving tools are consistently used. Full cross functional teams is/is not analysis, fish bones flow charts, and prioritization tools are effectively used
4. Analysis system	Structured problem solving analyses are supported with data and dates for problems occurrence, containment, root cause identification and verification, permanent corrective action, implementation and preventative action implementation. The structured problem solving tools utlizied includes a detailed verification plan which closes the loop on root cause resolution.
5. Advanced statistical methods	As per 4. and advanced statistical methods such as D.O.E.s are regularly used in team problems solving.

Figure 1. Problem-solving rating scale

Despite its title, as the extract above demonstrates, the Manual focuses on process, not quality. As Jim, the Human Resource Manager, says 'We could make life jackets out of concrete and that would be fine as long as we followed "the problem solving process"'. The *Autoco Quality Manual* does not specify what constitutes an appropriate standard (or 'quality') for the fabric that AFM produces, although specific contracts certainly do. The Manual is a way of controlling the way work practice is organized, the way people are to work together, and the things they do to make knowledge, to solve problems and innovate. If local practice can be aligned then integrating local practices into global supply chains should, the argument goes, be straightforward.

In order to score at least a 3, AFM needs to institute structured problem solving immediately, and it does so through the developing Eight Step Problem

Solving Teams. Eight Step Problem Solving Teams have to define, explore and solve a problem following a specific, written, template. The eight steps are:

Pinpoint–	State the specific area targeted for improvement
Process Definition–	How does the current process work? Flowchart if necessary.
People Supports–	(a) Critical People who will be involved in the XX??
	(b) Others in organization who will need some improvement
	(c) People who need feedback
Measure–	Define how we will track progress
Baseline–	Details of current performance
Goals–	Based on benchmarks establish final and shaping goals
Action Plan–	Define all steps and people responsible (*What*, by *When*, by *Whom*, *How*)
Feedback–	Specify how feedback will be delivered

For a problem to be deemed to be effectively solved, a team must be able to make an entry under each heading.

It is the process and purpose of forming a particular Eight Step Problem Solving Team that is under discussion in the Board Room on the Fluffy Floor. Participants are all members of the Action Learning Team, co-opted, in something of a hurry, by Jim. Baz was trained as a textile technician and, like Bill, has many years of experience in textile manufacturing, most of it at AFM. He is Production Supervisor and Bill's immediate Manager. Matt is relatively new to textile manufacturing, and to AFM, although experienced in other segments of the automotive industry. At the time of the meeting he is studying for his Bachelor of Business Administration degree. He is the newly appointed Systems Manager.

Bill and Baz and Matt and Sally and Margaret are adopting a team-based approach to problem solving, not because they are convinced of its effectiveness (although Sally and Margaret might be, Bill and Baz remain, as we shall see from the conversation, unconvinced), but because (as level 3 of the Rating Procedure above indicates) they are required to do so. They are using the 'structured problem solving tool' of the Eight Step Problem Solving Plan because they must do so in order to achieve an acceptable score and retain the custom of Autoco. As far as the *Quality Manual* is concerned, problems are only considered 'solved' if these processes and practices are enacted, and a written text is produced; other forms of problem solving do not register on Autoco's radar.

Problem solving is a particular pervasive form of knowledge production and is well documented in the business literature. It is unpredictable, hard to control and needs to be dealt with quickly 'on the ground' at the local site (or at several local sites). The Rating Procedure is an attempt to regulate what counts as knowledge at AFM by regulating the processes and relations by which knowledge is produced. The actions of Sally and Margaret and Bill and Baz and Matt, sitting around the table at AFM arguing about the purpose of team meetings, were put in motion by this document. The capacity of this document and others like it to authorize action and generate material texts like the Eight Step Problem Solving Plans rests on the conditions of its production as an organizationally warranted account of workplaces and work practices (Jackson 1996).

One of the features of written texts is what Smith calls their 'infinite replicability', and the *Autoco Quality Manual* certainly seems infinitely replicable, turning up in precisely the same format in South Africa at the Cape Town hydraulic factory and in Australia at AFM, demanding that precisely the same set of social actions count as knowledge production at every site. The *Autoco Quality Manual* is a good example of a material text that makes the global car possible. As Smith argues, it coordinates, orders, hooks up the activities of individuals in multiple local historical sites (: 149). Or at least, that's what it aims to do. The meeting is not going all that smoothly:

1	Sally	What you're saying has just prompted me to note down, just sort
2		of to think about, is, um, what is the prime purpose of these Eight
3		Step Teams? Now, I've interpreted it, and that maybe partly
4		influencing what's happening in Harborside as well, I've
5		interpreted it very much um that the team aspect of it. Now, if
6		you interpret it as: 'Let's get something fixed here' you may be
7		perfectly right, it may be [the]most efficient way to do that may
8		be to have you [individually] documenting what you do and
9		you'll come up with a solution.
10	Matt	yeah
11	Sally	Now, um, I think both, both things are
12	Baz	You get both sides, I think. Some plans you might have three or,
13		or five people on the team sort of thing, but in other cases, like
14		this case, Bill's the best person, IS the team
15	Sally	Yeah but if
16	Baz	or, if he wants me on there, if he wants me, but
17	Sally	But if there was a team around, um, Bill
18	Bill	No. It makes it too complicated because they always come up
19		with silly little things
20	Sally	But
21	Bill	Then, then you discuss, all that, and really you're going
22		backwards
23	Sally	Yeah

24	Bill	So to me it's a waste of the time we took. They tried this before
25	Matt	um Bill
26	Sally	yeah
27	Bill	and that's why it doesn't go far
28	Baz	in Bill's case I don't think the people have got the experience to do
29		those sort of things
30	Sally	What I'm wondering, though, is would they learn from Bill by
31		being involved in a team with Bill and seeing the way he goes
32		about it? If there'd be some spin off in terms of the
33	Bill	Not if you know, not if you know the people I work with. They
34		don't listen.
35	Baz	They don't listen
36	Sally	Which I don't [know the people you work with]
37	Bill	They don't listen, they like to go their own way.
38	Margaret	I think it'd be useful though Bill to try involving them in a team,
39		just to see what sort of experience they get out of it. What does
40		actually happen
41	Bill	But then I think you put the project back
42	Sally	But then we come back to what *is* the project? Is it to get people here working in teams or
43	Bill	no it
44	Sally	is it to solve problems?
45	Bill	Well, really, it's to solve problems. Make it a better working
46		place for the people, for the people that's here.
47	Sally	yeah
48	Bill	I mean they can't do it themselves so really it's up to me to
49		explain it to Margaret or whoever
50	Sally	yeah
51	Bill	What the costing is, which they don't know
52	Margaret	that was pretty much
53	Bill	and how to work it out
54	Margaret	I think that was the focus too
55	Sally	yeah
56	Margaret	when um it was launched by Pat and Jim. That we were looking
57		um at what problems can be solved
58	Sally	yeah
59	Bill	mm
60	Matt	Whereas in reality it's more of a as much a team building
61		exercise as it is as a dollar saving exercise or an improvement
62	Sally	exercise
63		It's an interest, it really makes an interesting contrast to see the
64		way both are going and I'm not sure that that this um question
65		has actually formed in the minds of the people who set this all
66		going, that that we would be getting this dual focus.

When Bill and Sally debate the purposes of the Eight Step Problem Solving Plan in the conversation above, I want to suggest that Bill and Baz are more focused on the 'problem' while Sally is more focused on the 'solving' and the 'plan'. In lines 1–9 Sally argues that they are equally concerned with 'getting something fixed' and 'the team aspect'. In line 9 Baz seems to agree, 'you get both sides I think' but he doesn't seem to mean that both aspects are equally important. Rather, he suggests that at different times in the workplace problems will demand different strategies–sometimes a team of several people but in other cases "Bill's the best person, IS the team" (line 14). Bill and Baz go on to collaboratively construct a model of knowledge production (or problem solving) based on 'experience' (line 28) to which other workers must first 'listen' (lines 33, 34, 35) if they are to make a contribution. Bill sees his role as to solve problems and so "make it a better place for the people who work here" (line 45–46) and to do that he must be the one who controls the knowledge, who 'explains' because 'they can't do it themselves'. It may be that, from Bill's perspective, it is his knowledge-producing skills which distinguish him from the other workers. Sally and Margaret construct knowledge as something that is collaboratively produced in a team, and Problem Solving Teams are teams in which people learn how to solve problems. Teams are an end in themselves. Significantly, however, although Sally is prepared to contemplate the novel notion of a team of one, all knowledge must be documented, "The most efficient way to do that may be to have you [individually] documenting what you do and you'll come up with a solution" (lines 7–9).

Their choices of focus are not innocent. Bill has a good deal invested in 'problems' being the focus of attention. The identification and solving of problems is part and parcel of the work of a supervisor, it involves expertise, real 'on the ground' knowledge of how things work in practice and the pragmatic constraints within which problems must be solved. For Bill, working knowledge is local knowledge; it comes with experience and invests the 'knower' with status. If Bill is the person who authorizes what counts as a problem and what counts as a solution (at least on the factory floor) then he effectively polices and legitimizes knowledge. Sally's position at AFM depends on her expertise in learning processes. She moves from industry to industry, developing only a superficial knowledge of what goes on at individual factory floors. Her expertise lies in learning theory—the 'solving' rather than the identifying of problems, and in developing written 'plans' like the Eight Step Problem Solving Plan the Action Learning Team is charged with developing. Sally's position within AFM relies on the reification of process, taking process out of the local context.

SOLVING PROBLEMS BY THE BOOK

After a break for sandwiches the conversation moves to a discussion of a particular Eight Step Plan, Bill's plan–the 'Buggies Project'. Bill, as a Warping Shed Supervisor, is concerned that warping is often delayed because his section has only one set of 'buggies', the structures on which the creels of yarn are loaded before being fitted to the warping machines. With the encouragement of the Action Learning Team, Bill has finally, and reluctantly, agreed to make 'The Buggies' the focus of an Eight Step Problem Solving Plan. Sally invites members of the team to speak about the progress of their Eight Step Plan teams, and Matt provides an explicit invitation to Bill to discuss his project:

1	Matt	I think, Bill, you've gone a fair way too on a couple of yours
2		haven't you?
3	Bill	Yeah. Me and Margaret, yeah, got together and I've worked out
4		quite a bit, lost downtime, and costing unloading
5	Sally	To what extent are you following the framework of that, that 8
6		step guideline that Jim [human resource manager] put out
7		originally in that, in the book where… Particularly step 7, where
8		its got the Action Plan and its got, um Who? When? Where? and
9		so on. Are you using those at all?
10	Bill	No, I'm just using this one [points to his head].
11	Sally	yeah [Margaret]
12	Baz	[ ]
13		(laughter)
14	Matt	Baz filled one in.
15	Sally	Ah, yeah.
16		So you are following that paperwork there?
17	Bill	Yeah. I don't know. Are you Margaret?
18		[loud laughter]
19	Sally	I'm sorry, I've asked the wrong person
20	Bill	I'm just writing it all out and giving it to Margaret. That's all.
21	Margaret	I'll give you an example. This is one that we are working on,
22		that, um, Bill's talking about.
23	Sally	Yeah
24	Margaret	And these are really all his figures.
25	Sally	Oh, I see.
26	Margaret	And so, when we get together, that's the sort of thing we do.
27		Writing up all these things but
28	Sally	Oh, OK.
29	Margaret	then gradually translating it on to here
30		[Margaret has a pile of three folders]
31	Matt	Matt [We're] sort of saying that Margaret's an excellent
32		facilitator [laughter]

33	Sally)	and is, is there actually a group involved with doing that or
34		[yourself]?
35	Bill)	No, because I find, I find it's too awkward for Margaret if she's
36		talking to individual people, people that's only been there nine
37		months,
38	Sally	yeah
39	Bill	so they don't know [what] they're talking about half the time
40		and they're giving Margaret the wrong information. So I said to
41		Margaret, it's better if she comes to me and I can give her the
42		right times and all this you know.
43	Sally	Yep Yep
44	Margaret	But what we talked about last time, though, was that it really
45		had to be set down, so what I've done is write down
46	Bill	Yeah yeah
47	Margaret	your name and Nick who had been involved in talking about
48		some of it too and Bashir
49	Bill	yeah.
50	Margaret	now I did have Yuan but you, you felt that he's not in that area
51		now
52	Bill	no
53	Margaret	so you prefer to have Chris.
54	Bill	Yeah
55	Margaret	And Matt, your name was now put on to that one as well
56	Baz	[growls]
57	Margaret	So that's a bit, a bit of a way that things are operating though, so
58		we haven't really started off with a team that's taking
59		responsibility for it, so I guess, in a way, it's um evolving as it
60		goes.
61	Sally	Well, um, mm yes.

What I want to do here is to focus on the way that individually and as a group the Action Learning Team integrates a particular material text into the practices and social and historical relations of their local workplace. I'll start by presenting a general reading of the transcript and follow that discussion by focusing on a single word to explore the way in which knowledge, identity and workplace relationships are absolutely central to the way the written text of the Quality Manual is integrated into the local site.

The transcript begins with Matt inviting Bill to describe the Buggies Project. Matt identifies Bill as the 'owner' of successful Eight Step Plans:

you've gone a fair way too on a couple of yours.

Bill does not take up the invitation to describe his work on the Buggies Project as an Eight Step Plan and himself as the leader of an Eight Step Plan Team. Instead he calls on traditional work discourses to describe his work on the Buggies Project as practice, something that is performed rather than written: '[we] got together', 'I worked it out'. He clearly identifies himself as the 'primary knower' (Berry 1981) who performs the action of '[working] it all out' (producing knowledge) while Margaret, the literacy worker who is later assigned the secondary task of writing it down, is not mentioned at all at this point.

Sally uses her turn (line 4) to reassert the primacy of textualized knowledge as the only acceptable way of producing knowledge in the meeting. Framed as a formal question reminiscent of the literate practices of schooling ('To what extent . . . ') she offers no obvious space for resistance. Her focus is exclusively on the textual properties of the Eight Step Plan. It seems that Bill's experience must be framed within the protocols of Action Learning if Sally is to accept it as authentic knowledge production. Sally specifies the mandated-text structure: 'Who?', 'When?' 'Where?' (line 8) and calls on institutional authority (Jim) to reinforce her own ambiguous authority as a low-status female literacy worker and high-status expert outsider:

'that eight step guideline that Jim (the Human Resource Manager) put out' (line 6)

Bill explicitly rejects this offer to express his knowledge in the abstract and textualized framework required by Action Learning frameworks. He locates his knowledge 'in the body' rather than in the word. In pointing to his head (line 10) Bill interrupts Sally's attempt to textualize his knowledge by referring to his knowledge as embodied. In doing so he reasserts traditional work discourses in which knowledge is embodied practice and personal experience is the authority that counts.

Matt's turn (line 14) continues to distance Bill from the requirement of textualizing his knowledge. In saying that 'Baz filled one in' he acknowledges the formal requirement to document the process while reducing its significance by implying that it is familiar and routine—just another form. Sally takes up Matt's lead, referring to the Eight Step Plan documentation as 'paperwork' and thus locating it away from the 'new work order' and in traditional workplace discourses in which paperwork is regarded as a trivial irritant.

1	Matt	Baz filled one in.
2	Sally	Ah, yeah.
3		So you are following that paperwork there?
4	Bill	Yeah. I don't know. Are you Margaret? [loud laughter]
5	Sally	I'm sorry, I've asked the wrong person
6	Bill	I'm just writing it all out and giving it to Margaret. That's all.

Bill reinforces the familiar roles assumed by himself and Margaret in their cooperative work. He uses his head. Although he does '[write] it all out' (line 20), he then 'gives it to Margaret' and Margaret is responsible for whatever formal documentation is required. Matt, Bill and, for the moment, Sally, have cooperated in uncoupling knowledge production and text production, preserving Bill's authority while collaboratively constructing a working identity which does not include writing. Poynton (1993) and Noon and Blyton (1997) have demonstrated that, in traditional workplace discourses embodied knowledge is privileged in relation to textualized knowledge and that literacy practice is regarded as the particular domain of women and of secondary value. Here Bill has reasserted traditional work order discourses, and in doing so he has reasserted the systems of value and belief, and the working identities, which are embedded in it. He has transformed the Buggies Project into a workplace problem which needs a practical solution, rather than an Eight Step Problem Solving Plan in which documented process is paramount.

Like Matt, Margaret uses her turn to resituate the Buggies Project in the discourse of Action Learning while not explicitly contradicting Bill's location of the activity in traditional industrial and gender relations. First, she acknowledges Bill's working knowledge 'these are really all his figures' (line 24), but she says that when they get together 'that's the sort of thing we do. Writing up all these things' (lines 24–25). She identifies the writing task as a translation task—translating knowledge from the head to the word. Matt specifically states that this is the task of the workplace educator 'Margaret is an excellent facilitator', but Sally queries this attempt to reinterpret the meetings between Margaret and Bill as a collective construction of knowledge, asking directly 'is there a group doing this?', and Bill locates himself once again as the primary knower:

I find . . . it's too awkward for Margaret (line 35).

He rewords 'the group' identified by Sally as individual people who 'don't know what they are talking about half the time' (line 39) and reasserts the private and hierarchical nature of his relationships with Margaret:

so I said it was better if she comes to me (line 41).

Margaret resists this notion, reclaiming the force of documentation, 'it really had to be set down so what I've done is write it down' (lines 44–45) to make the knowledge available to the group. Her final comment (lines 57–60) reframes the experience of the Buggies Project as a process 'it is evolving as it goes' which locates it again, although precariously, in the new work order discourse. Sally, the strongest voice of the new work order discourse, seems to have accepted that she has lost this round.

I want to turn now to the micropolitics of the text, focusing specifically on the difference between the active and the passive voice. The segment of the team meeting that is presented here represents a particularly acute moment in a struggle over knowledge production. The importance of the moment is signaled when Bill uses the term 'costing' (line 4). 'Costing' is a nominalization (Norman Fairclough 1992: 27). A nominalization transforms 'I worked out how much it cost' into 'the costing is'. The salient feature here is that nominalization obscures agency—the person who produced the knowledge is stripped from the text. In this case the term 'costing' is used by Matt and others to refer to the written account of what a particular process, plan or innovation will cost. It usually refers to a written description in which the calculations are spelled out in routine ways, but it may also be used to refer to calculations as if they existed in a routine documentary form (what would the costing be for that?). It is a term used most commonly in management discourses drawing on accounting and systems management so it is not surprising that it was introduced by Matt, in an earlier meeting. It is however, a term rarely heard on the shop floor, until recently. It is not that people on the shop floor did not perform calculations, or present those calculations to their supervisors. At AFM at least, people in Bill's position have always done that, and Bill has taken pride in his ability to make a clear financial argument to his superiors. In this transcript Bill uses a number of active formulations:

> I've worked out quite a bit (lines 3–4)
> I'm using this one (line 10)
> I'm just writing it all out (line 20)
> I find it's too awkward for Margaret (line 35)
> I can give her the right times (line 42)

In each case he stresses his individual agency by placing himself first in the clause (I) as well as stating it explicitly in lines 39–42. Nominalization does not come easily to Bill, yet nominalization is a critical feature of the discourse of collaborative learning and of the Action Learning team. This is no accident nominalization reassigns knowledge away from individuals to groups or unidentified individuals.

Bill comes from a sociohistorical tradition in which individuals know, and groups do not know. For Bill, working knowledge is embodied knowledge; even abstract knowledge is located in people's heads, not in documents. Working identities are directly associated with having this knowledge. Within this tradition Bill occupies a privileged position. His knowledge, firmly located in his experience, invests him with power, status and authority. It confirms his gender identity. Embodied knowledge is the kind of knowledge that has always counted and textual knowledge is of secondary importance, indeed, it is associated with women who have secondary status. Bill uses nominalizations of knowledge processes (like 'costings') awkwardly, preferring formulation like 'I worked it out' in which the knower is clearly identified. Sally and Margaret, on the other hand, come from a sociohistorical tradition in which textual knowledge is the knowledge that counts. Within this discourse Sally and Margaret hold privileged positions. Their knowledge invests them with power, status and authority. The documentation of process is paramount and embodied knowledge only comes to count when it has been textualized. A feature of this discourse is the nominalization of process (like costing); a procedure which renders the knower invisible and so challenges traditional working identities and hierarchical relationships.

Talk and text

When people talk about the ways that ICTs affect workplaces they often concentrate on numbers—are there more texts? Are there fewer texts? Are images now more important than words? Texts are critical to globalization; they are critical to making knowledge common, in so far as it can be, across global webs of production. In this chapter, however, I've argued the really important questions aren't about the quantity of texts, or about the distinctive characteristics of the texts. The critical questions are 'What work does the new text do? How does it impact on the social relations of work? How does it change the power dynamic? How does it change relations of power and authority? How does it change what counts as knowledge at the local site? What about remote sites? What resources do people use to interpret texts? How are the texts integrated into local work practice?' As we have seen in this chapter, new written texts like the *Autoco Quality Manual* generate new social practices at work, even for people like Bill, who have no direct contact with global corporations. They also generate new texts in their turn. They reconfigure existing relationships and they recalibrate power relationships.

Many people argue that the texts and technologies of the electronic workplace obviate the need for social interaction, that they regulate and standardize practice to the extent that face-to-face communication is unnecessary and print is redundant. This may happen sometimes but, equally, the electronic workplace demands new and different kinds of face-to-face interaction, and generates new and different kinds of print texts. The task of integrating new texts into local workplaces, and using them to join up global workspaces, is complex and is achieved collaboratively. New face-to-face practices like the Action Learning Team meeting grow up around the new texts. Social relationships are reworked through those social interactions. The print text changes the way that social relationships are regulated, the way authority is exercised, the way work practice is represented and the way knowledge is produced and legitimated in a local workplace. It generates new communities of practices in the local workplace—not only in global workspaces. The print text is one of the ways that ICTs affect, 'even in the smallest way', local practice.

These changes are, however, reassuringly unpredictable. This is because no one can know the context in which the new text will be animated in the local workplace—they can't know the power relationships that exist, the intricate work practices that have evolved over time. Global corporations want to actively exploit the unique qualities of local workplaces—they don't want all workplaces to be the same. But they do want standardization of work practices and control and surveillance. It is workers like Bill and Baz and Matt, operating in local workplaces like the Exmouth Plant and the Harbor Mill, who integrate texts like the *Autoco Quality Manual* into their work practices, and who produce texts like the Eight Step Problem Solving Plans.

Within this context, Sally and Margaret's roles as educators are also undergoing uncomfortable reconstruction. While they have a new, and far more central role in the organization, they are inserting themselves into very well-established work practices and work relationships. Margaret's role with Bill's Buggies Project, for instance, is reminiscent of that of a secretary—Bill dictates and Margaret transcribes. It is difficult for her to resist this division of labor and still achieve the outcome that AFM requires. In this particular struggle over 'what counts' as knowledge at AFM, gender roles play a part and are invoked when traditional formulations of knowledge are challenged. Some of the dilemmas faced by Sally, Margaret and other workplace educators are taken up in Chapter Six.

This chapter focused on a relatively conventional way in which knowledge which is used globally is produced textually in local contexts. Phenomena like the world car demand that people all over the world communicate more immediately and directly than this. In the next chapter I turn to the textual practice of virtual teams solving problems in cyber-place.

CHAPTER FIVE

Solving problems in cyber-place

> The development of new communications infrastructures is
> not some value-neutral, technologically pure process, but an
> asymmetrical social struggle to gain and maintain social power,
> the power to control space and social processes over distance
> (GRAHAM)

INTRODUCTION

In Chapter Four I talked about the way a print document shifted the microprocesses of working life at the Exmouth Plant and how workers at the Exmouth Plant collaborated to reinterpret the print document in the light of social and historical context into which it was integrated. An electronic text, transformed into a hard-copy print text, changed the way that people talked and worked, changed the way that knowledge was understood, produced and legitimated, and changed the power relations of the workplace, in both subtle and obvious ways. In this chapter I want to look at on-line environments, and the ways they emerge from, and are integrated into, local workplaces. The on-line environments I am referring to are exclusively textual environments in so far as they appear to rely largely on digitized print and graphics to generate knowledge and solve problems. They are frequently presented as comprehensive, global, problem-solving environments which exploit the possibilities of communications technologies removed from sordid human behavior. There seems to be something elevated and noble about problem solving in cyber-place, where knowledge production is uncorrupted by the petty social and political histories of local factories and offices.

It is not my intention to diminish the productive potential of on-line environments in this chapter. Rather, my aim is to demonstrate that the best use of these environments can be made when we recognize that they are material and ideational spaces like any other. They have the same kinds of histories as any physical workspace (although these histories are likely to be more complicated and even more likely to be taken for granted), and they call on workers to use their existing resources and to improvise new textual practices to make them work. As with any workspace, on-line environments invoke complex power relations, and they stimulate power struggles. The struggles and compromises of on-line environments will percolate into various local workplaces, changing work practices and being changed by them. If we wish to understand how to learn and teach the textual practices of on-line environments at work, we need to understand how they are integrated into existing practices and processes of knowledge building, and how resources are mobilized.

We also need to understand the complex ways in which people read on-line environments and situate themselves within and between them. Relations across on-line environments are social as well as technical, and people and texts create the join between the on-line environment and the local physical workplace—they are the place where, as William J. Mitchell (1994) puts it, 'bit meets body'. A feature of on-line communication is that the centers of cities can be more closely connected to each other than they are to their own peripheries. It is often left to workers at specific workplaces to interpret the working of the on-line community to the periphery, and to make the critical local knowledge bases of the periphery available to the problem-solving on-line community. This is a complex process of interpretation and translation.

In this chapter I focus on an on-line environment that forms part of the potential workspace of Bill and Matt and Baz and Grace and Mary. First I describe the kinds of interventions that communications technologies can make in local workplaces, then I describe a particular on-line workspace, from a social rather than a technical perspective. Finally, I look at the kinds of issues it raises for the people who work at the Exmouth Plant.

CREATING TEAMS ACROSS SPACE AND TIME

When Bill and Matt and Margaret and Sally sit around the table to work through the Buggies Project they are cobbling together new work practices from the practices that have served them in the past, from experiences they have had elsewhere and from the practices of a transnational corporation as they are described (and no doubt idealized) in the *Autoco Quality Manual*. Although it may not seem obvious to anyone who passes by, they are engaged in one of the routine microprocesses of globalization. They are collaborating in order to make a globalizing print text work. Without their individual and collective efforts this particular instance of globalization wouldn't happen, and AFM would be at risk of losing its coveted place in Autoco's global network of suppliers.

Reading and writing are central to this globalizing process. Unfamiliar texts like the *Autoco Quality Manual* have to be pushed and prodded, modified and adapted to make some kind of sense to the diverse group of people sitting around the Board Room of the Exmouth Plant. New texts, like the Eight Step Problem Solving Plans, have to be imagined and written by those same people in ways that are likely to satisfy the mysterious, invisible readers on the other side of the world at Autoco Head Office. This is not an easy task, and it requires inventiveness, sensitive negotiation and a fair degree of compromise. To the extent that it actually happens, it happens because the team members talk their way to a more or less agreed-upon position of what has to be done, how it will be done, and, critically, who will do it.

But talking, reading and writing are fairly conventional activities in workplaces and the influence of globalization on the everyday work of the team may not be immediately obvious to the casual observer or, indeed, to Bill and Matt and Sally and Margaret, and the others. They are, after all, sitting together in the same room, talking about familiar physical environments and familiar work processes. If they want to, they can all troop down to the Factory Floor and view the buggies and the machines, watch the men load the creels, see the skill and the physical exertion the job requires, and get an idea of the problem as Bill and the men on the floor see it. So, while they are all operating in a global network of communication, up close they look (and perhaps feel) like a self-sufficient unit, and their connection to other nodes of the network—like the workers who are puzzling over the same *Autoco Quality Manual* while they put together hydraulic components in South Africa, for instance—is obscured.

For other people, creating workgroups across space and time is a more intimate and intrusive experience. Communications technologies have become familiar material objects in their workplaces, so the globalization of their work is difficult to ignore. In the Australian trading room of a global energy company, for instance, a huge video wall transmits larger-than-life size images of workers in Hong Kong drinking coffee, chatting with each other, talking on the phone, and entering data. Like their colleagues in Australia, they are trading futures on a global energy grid, and their virtual presence in the trading room is now, literally and figuratively, just part of the scenery. In theory, at least, people on either end of the video link can start a conversation, or join one, at any place and any time—the idea is to replicate the routine relationships of a face-to-face workgroup as nearly as possible by simulating physical presence. In this case at least, the visibility of the remote team members doesn't translate into easy office-to-office banter. In practice the mute button is nearly always activated, and people rarely glance up to the video wall unless they are called to attend to a particular problem. They use E-mail and the telephone to communicate with each other.

Other workgroups are held together by communication webs of more familiar and pervasive technologies. Written documents, mobile and landline telephones, text messages, pagers, video and voice conference links, desk top, laptop and palm-top computers and, of course, E-mail, connect people who are located hundreds, even thousands of kilometers and many time zones, apart. Individuals may, at various times, access their messages, send and receive data and documents, and conduct their conversations in airports and cars, coffee shops, offices and their homes at any time of the day or night—indeed they may be required to do so if they are to have any kind of regular contact with their colleagues. Within these contexts people often have favored channels of communication, and they make use of different communication channels, or different combinations of channels, depending on the person they are communicating with and the subject they are communicating about. The multiplicity of communication channels, and the sheer volume of text-based communication that they generate, can be both a burden and an opportunity. At PaceSetters (Farrell and Holkner 2003), a global company manufacturing and selling medical appliances, communications technologies are linked together through a sophisticated technical protocol (Bluetooth) so that workers, wherever they are, can reach each other and access important technical and legal documents located in shared databases on a staff intranet. Technicians operating 'in the field' in hospitals and medical offices keep contact with colleagues at the local office, team members in offices in other states, and colleagues in other countries through a complex and informal protocol of phone, paper memos, E-mail, pager and text messages. There is no shortage of ways to keep in touch. The challenge is to avoid being buried in an avalanche of text coming through every communication channel at all hours of the day and night. An E-mail message may be a report of an important clinical trial, an invitation to join the football tipping competition, a request to provide feedback for the marketing team, an important message from the US-based Head Office or a joke someone thought was funny enough to send as a broadcast message. Individuals and teams need to find ways to navigate through this mountain of documentation.

One group of technicians at PaceSetters who operate in one geographical area and rely on each other heavily for advice and support has devised a code for messages between themselves, indicating the level of urgency. A '1' asks for an immediate response; they may be assisting at an implant for instance, and, with the patient on the operating table, be experiencing trouble programming the appliance. A '2' asks for a response within 30 minutes, a '3' within a couple of hours, a '4' by the next day, and so on. They prioritize pager, SMS and voicemail messages from each other, and attend to other channels 'when they have time', which may not be often. They are renowned for ignoring E-mail messages for days at a time (they almost never use E-mail with each other) and have been known to neglect to reply to E-mail messages from their branch office for up to four months. While multiple channels of communication generate a volume of text that is impossible to deal with and could be experienced as a burden, the technicians use it as an opportunity (or an excuse) to claim some agency in their working lives and to exert some control over who and what they attend to, and when.

Still other workgroups, especially those operating in small businesses, supplement routine face-to-face and one-to-one telephone contact with computer-mediated communication when specific circumstances seem to need it. Edubase is a small family-run business developing software for education markets all over the world. While members of the workgroup maintain close personal and business ties through face-to-face contact, telephone and E-mail, when a group meeting is necessary they co-opt generic, freely available software (hotmail instant messaging link ups for instance) to supplement their normal communication channels.

What is distinctive about these kinds of workgroups, and what makes the development of collaborative problem-solving processes possible within them, is that while they are geographically distributed, the work group itself is stable and on-going, it has multiple communication channels available to it, and it makes choices about how and when to use them. The new technologies are supported by established technologies and new technologies provide occasions for using older, established technologies (landline telephones and written documents for instance) in new ways. The manner in which the technology is used evolves from the existing work practices of the group, and these practices evolve as the new technology is established. While they are certainly heavily dependent on written text, work groups like these rarely confine themselves to a single channel of communication.

But what happens when a globally distributed work group of people is established for a specific project, relies on one channel of communication, has no process for integrating into local workplaces, and no process for modifying or adapting the channel of communication as the group evolves? What happens when a team exists exclusively in cyber-space?

A DAY IN THE LIFE WITH COVISINT COLLABORATION MANAGER . . .

> During the product development process, a vehicle team receives direction from marketing to increase the capacity of a glove box. The engineering lead for the instrument panel is given the task to incorporate a larger glove box while maintaining functionality and performance. The engineer learns that the supplier for the glove box door needs to finalize the tool in three weeks to support the vehicle launch deadline.
>
> The modification requires coordination throughout the vehicle program team (engineering, interior styling, human packaging, testing) as well as the supplier's organization. The engineering lead estimates the change will take six weeks to implement with their current process - three weeks longer than the supplier has for final tool modification.
>
> The engineering lead decides to use the Covisint Collaboration Manager tool. He logs on to Covisint and creates a workspace for the Glove Box Modification project. Then, he adds team members, loads the marketing specification documents, and the project-specific documents, into the workspace. The team members are notified via e-mail and receive a URL directing them to their workspace.
>
> In the next three weeks, the team uses Collaboration Manager to review documents, conduct virtual design reviews, and assign and track issues. The team quickly reviews all design decisions and carries out various "what-if" studies. Through the use of Collaboration Manager, the designer develops a solution for the glove box and uploads the new design on the workspace. The design is approved and the supplier's tool deadline is met. (covisint.com/solutions/collab/res/adil_collab_mgr.shtml 14/01/01)

If the Collaboration Manager is the solution, what's the problem? From Covisint's point of view there are two, intimately connected, problems. The first problem is a simple one—it is concerned with the ready accessibility of data and information. Critical people, and critical information, are dispersed across space and time. When critical people operate collaboratively in virtual workspaces they need a stable information base on which to rely as they make business decisions as a team. Different operations are located at specific sites for good reasons, but when people, departments and suppliers need to communicate precisely and under pressure, distance becomes a problem. There is a pressing need to coordinate activity in different locations and time zones and, especially, to make sure that everyone is working from the same (rapidly changing) information:

> With the industry shift to outsourced engineering, program teams have become widely dispersed groups composed of members from different companies and geographic regions. In addition, product development cycles are shrinking. As a result, team members must have access to information to execute business decisions quickly. To co-ordinate virtual teams and make critical program information readily available, team members need to be able to collaborate effectively. The team needs one central source of information on which to base their daily business decisions. (covisint.com/solutions/collab/collab_mgr.shtml 22/11/01 12.31pm)

However, while there certainly is a problem around space, time and information, there is also a problem about the nature of the social action generated in spatially and temporally dispersed teams. What really slows down production is 'friction':

> The communication of information between the trading partners is riddled with friction. To remove the friction from the sourcing process, buyers and potential suppliers must engage in a collaborative process. (Covisint on-line)

So, the Collaboration Manager is designed to solve the problem of 'friction' by simplifying and streamlining the collaboration process. It mandates one channel of communication, a highly regulated electronic workspace controlled by the team leader. If the trading partners are prepared to enter the electronic workspace they will have access to the same information at the same time and, it is implied, the friction will be removed. (An incidental outcome is that there will be a durable text record of the deliberations, processes and solutions the team develops that can be retained by the company and, in theory at least, be used for the next project. Project teams are notoriously bad at transferring knowledge and processes from one project to the next, and knowledge is 'sticky'; it tends to stay with the people who make it.)

The problem of friction in short-term project teams cannot be reduced to the straightforward matter of providing access to shared databases; friction is a feature of all workplaces, and probably all human activity. Very often project teams are essentially groups of people located in different countries, companies and professions, brought together to solve a single identifiable problem that involves them all, like redesigning the glove box on the run. These people may never have worked together before, may not do so again, and are unlikely to meet face to face, or to recognize each other if they do. The people, the companies, even the countries, may have a history of conflict that will be difficult to put aside for the duration of a single project. They have played no part in developing the mandated collaboration processes, and they are relying on a single communication channel—a computer software program–to communicate everything that needs communicating to a punishing time line.

This chapter looks at the challenges facing project-based virtual teams, and what Bill, Matt and the others can teach us about what might be involved in solving problems in cyber-place.

VIRTUAL WORKSPACES/VIRTUAL WORKPLACES

One of the most seductive aspects of the virtual workspace is that it gives the impression of being a blank screen, a Greenfield site on which participants can engage as equals (without gender, class, race, creed or generation to distinguish them) in the disembodied, disinterested, intellectual activity of problem solving. The attraction of the virtual workspace seems to be that it removes physical, geographic, temporal (and even sometimes linguistic) barriers to direct negotiations between people and organizations. People operating within virtual workspaces like the one described in 'A day in the life . . .' are invited to think of the workspace as existing 'in the ether', as a stream of digital impulses pushing raw data from one electronic location to another, unsullied by human contact. It can be easy to forget that a virtual workspace is made up of 'things' like satellite dishes and optic fibers that come from somewhere, and are inserted somewhere, in particular. It can be easy to forget that a virtual workspace is created by people and institutions with their own histories, politics, relationships and agendas.

The virtual workspace is not a neutral space inhabited by avatars engaged in the purely cognitive activity of problem solving unhampered by the corporeal, cultural and ideational constraints of physical workspaces. Neither is it an unmediated environment for the direct exchange of information. A virtual workspace like the Collaboration Manager is a built environment like any other. In important respects it is just like a physical workplace; it came into being because of particular social conditions and economic and historical events. Decisions about where it is located, the building blocks it uses as its programming foundations and the structural and decorative elements of its 'architecture' are at least as political as they are technical. Technology is talked about as an abstract system, existing in splendid isolation, but it is never abstract when it is enacted.

Once it is constructed, all activity in a virtual workspace is shaped and constrained by, and to some extent interpreted through, the 'architecture' of the site. There is, however, a good deal to be gained by normalizing the structure and practices of a virtual workspace, encouraging people working in a highly charged environment to believe that here, at least, is a neutral workspace where they can deal directly with data, information and people, where they can be themselves—or at least be who they want to be. However, if we pretend that virtual workspaces are neutral constructions we will never understand what people have to do to operate effectively in global workspaces.

It is also dangerous to treat virtual workspaces as if they are, in fact, entirely virtual. Virtual workspaces are always integrated into networks of local, physical workplaces like the Exmouth Plant and the Harbor Mill; they have to be if they are to effectively 'join up' remote global trading partners. When a member of the glove box team can't make the program work, or can't find the information she needs, or can't make himself understood, they will more than likely call on their immediate colleagues, the IT whiz in the office next door, the production manager who has access to all the data, the mate down the corridor. New technologies do not necessarily obviate the need for social interaction; they are just as likely to create new opportunities for local face-to-face interactions of the problem-solving kind. If they are to 'work', virtual workspaces have to be woven into the fabric of existing physical workplaces where people sit at their computers, entering the virtual workspace for a while and making the texts that create the join between the local and the global.

THE VIRTUAL WORKSPACE PROJECT

The Collaboration Manager is part of the Virtual Workspace Project; a suite of web-based applications (including a 'Problem Solver' and a 'Quote Manager') developed for the automotive industry by a company specializing in e-commerce. A little digging to uncover the archaeology of the site, however, reveals an epic tale of power, glory, pride and falls.

The Virtual Workspace Project was funded by a consortium of three dominant automotive manufacturers—General Motors Holden, Ford Motor Company and Daimler Chrysler. The aim was to build a giant electronic 'B2B' (business-to-business) exchange to streamline the production of the global car in a network of production that extends throughout the even moderately industrialized world. Since the Second World War the automotive industry has dominated global economic activity, and, at its inception (in 2000), the project was expected to be by far the largest and most ambitious e-commerce enterprise in a climate in which large and ambitious dot.com activities were the norm. Existing, much smaller, B2B exchanges in the automotive industry expected to be swamped by it if they were not able to secure a piece of the action for themselves.

The initial impetus for the project was to discipline the automotive companies' suppliers. Each automotive company was dealing with hundreds of different supplier companies (manufacturers of gearboxes and distributor caps, upholstery fabric and windscreens, air-conditioning systems and body trim, and they were quoting for thousands of specific jobs in various parts of the world under different environmental and safety standards and specifications. And, as the design of a particular car evolved, the standards relating to each component changed. So, when the glove box-design changed, the design specifications for the plastic mold had to change too, and new quotes sought. Automotive companies were under pressure to choose between quotes that were not commensurate and to deal with supplier companies whose systems were not compatible. So, one important aim was to have various suppliers (trading partners) standardize parts and their methods and templates for quoting on jobs, so that prices and contracts could be compared directly and updated automatically as the design of a new model car evolved:

> The quotation process is ideal for this type of improvement, according to Swift [CEO of Covisint]. The old-fashioned way is extremely inefficient quote comes through, gets faxed six times, and no one can read it, for example. Or the design alters, and only five of the eight quotes get updated. "When it's e-enabled, it's instantaneous. You move at the speed of light, which means you can update a quote with a supplier immediately to lay in design changes, so they make better decisions faster. That is going to be a huge benefit for the industry." (Butters and Bennett 2002 online)

Another and far more complex aim was to provide a systematic way of coordinating the various contributions to the complicated process by which a new design is brought to production:

> "If I'm sitting in Cologne, Germany, and the part is being designed in Hiroshima but will end up in Kansas City, how do I get all the constituents together—engineers, manufacturing people, suppliers, and purchasing people—to agree on how this part is going to progress?" asked Swift. Instead of relying on faxes and difficult-to-set-up global meetings, the software allows people to communicate instantaneously using a collaborative interface (Butters and Bennett 2002 on-line).

The Collaboration Manager is a software 'tool' designed to address this specific problem. It provides a common electronic workspace where people involved in various aspects of the design process work together to solve problems under severe time pressure. The hybrid technology of the fax, the phone, the courier and the mail are presented as failing to meet the time frame when an organization has to move 'at the speed of light'. The problem here is again represented as essentially technical—a matter of creating an interface to get everyone together. The implication is that, once together, getting everyone to 'agree on how this part is going to progress' will be a simple matter. The story of Covisint as it is documented through the financial and industry press suggests that treating the virtual workspace as a simple technological accomplishment was a mistake.

THE VIRTUAL WORKSPACE PROJECT

Covisint (combining, optimistically as it turned out, 'cooperation' and 'vision') was established in 2000 by a consortium of the 'big three' automotive manufacturers. It was an unlikely alliance, given that the level of competition between the companies was described in the industry press as 'bitter enmity'. The aim was to cut costs. This was to be achieved first by forcing suppliers to standardize their products and business processes to those preferred by the major automotive manufacturing companies, and second by providing a set of templates (virtual workspaces) to facilitate and regulate direct communication and cooperation between the major trading partners, thus saving time (for the automotive manufacturers, if not the suppliers) and, therefore, money.

The 'bitter enmity' that had characterized the relationships between the big three automotive manufacturers was not put aside in the development of the Virtual Workspace Project. It took them three months to agree on a name. Each company had elements of an electronic 'B2B' exchange, customized to their precise needs, in place (though nothing on such a comprehensive scale imagined for Covisint), and each worked with their preferred software development company. Each wanted their own company's template to become the prototype for the relevant application, the model used in the Virtual Workspace Project. Naturally enough, each of the software development companies wanted a stake in this huge project, the alternative appeared to be annihilation:

> Not surprisingly for a Net project, technology conflicts arose almost immediately. Ford had settled on Oracle for software to run its exchange, while GM had chosen upstart Commerce One's (CMRC) marketplace software. Although they're

supposed to work together, "they haven't gotten down to how they're going to execute all of this," says Carl Lenz, research director with Gartner Group Inc. Then DaimlerChrysler began asking why SAP's software, used in its plants, couldn't be employed (Welch 2000 on-line).

The design of the Virtual Workspace Project was an uneasy compromise between these interests. The impetus for the Virtual Workspace Project came in significant part from the intensifying competition and the global reach of the automotive industry in the global economy. With the cacophony of institutional voices rivaling the Tower of Babel it seemed pragmatic to cooperate to standardize communication (of data, information, negotiation and problem solving) across the industry. The competing institutional interests and the institutional histories of antagonism between the auto companies worked against a purely pragmatic, technical approach to the problem. This was not the only constraint on design decisions, it seems that even the history of personal relationships between key personnel came into it:

> To make matters worse, Commerce One Chairman Mark B. Hoffman and Oracle Corp. boss Lawrence J. Ellison have been dire enemies ever since Hoffman ran Sybase Inc., which nearly capsized trying to battle Oracle (Welch 2000 on-line).

So, when the technicians came to design the Virtual Workspace Project, they were already constrained by negotiations, compromises and decisions that were political and historical rather than technical in nature. For the technicians, the task was to interpret the communication needs of the B2B exchange in the light of the tasks the computer could actually be made to perform. This meant that in defining what happened in quoting, or problem solving, any task that the computer couldn't perform (no matter how crucial it was to the process) became invisible. The Virtual Workspace was not meant to provide a partial contribution to problem solving or quote managing; it needed to appear absolutely comprehensive, to leave no gaps in the process, if it was to justify the huge amount of money, time, and emotional labor, spent on it.

The problem was exacerbated because it is an almost impossible job to exhaustively describe even the simplest work practice or work process; so much that seems to be incidental local knowledge—like ignoring an out-of-date question on a routine form—turns out to be integral to the efficient flow of work. And, while there are certainly inefficient ways to do things, there is not one best way to perform any of these functions, and certainly not for contexts as diverse as a Cape Town hydraulics factory and a Melbourne upholstery manufacturer. The technicians relied, as instructed, on the 'big three' automotive companies to get a sense of what was needed in the virtual workspace. The applications they developed reflected the automotive companies' perspective on what constituted 'a problem' and what was needed to get a problem solved, as well as the possibilities and limitations of the technology. The suppliers, who were supposed to use the virtual workspace, had no input in the design.

Of course, the success of the Virtual Workspace Project depended on the participation of the major trading partners, the suppliers of gearboxes and engines and distributor caps and seats and upholstery, and many of them threatened to refuse to use it. It seems that the big three had overestimated their coercive power with their major suppliers. Suppliers were not persuaded that the Virtual Workspace was neutral ground. From their point of view, the giant automotive companies had a history of relentlessly driving down the costs of parts, and they did not see why they should cooperate in a process which was likely to put them at further disadvantage. While the automotive companies expected to reduce their own costs, suppliers would attract all the costs of setting up the new business processes required by the Virtual Workspace. Since they had no say in developing the processes of the Virtual Workspace, and since the processes were designed to suit the automotive manufacturers rather than the suppliers, integrating local work practices with the Virtual Workspace was likely to be difficult, time-consuming and expensive and possibly ultimately unsatisfactory. Traditional competition (sometimes described in the industry press as 'hostility') between the suppliers exacerbated the problem. Suppliers were suspicious that the car companies were not trustworthy, and since they controlled the virtual workspace, it was unlikely to be secure and would make confidential information available to competitors, who would then under quote. In general they did not have a history of 'collaborating' directly with their competitors, and they did not trust, or even understand, the alien processes set up to produce a collaborative environment.

In short, the Collaboration Manager, which is presented as a neutral environment for collaboration and problem solving, is more like a battlefield than a Greenfield site, even before people like Bill and Matt start using it! A chastened CEO of Covisint reflected that:

> We spent a lot of time developing products from the voice of the car manufacturer and didn't spend enough time getting input from the suppliers. I think you've got to understand who your customer base is, and you need strong input relative to what products you are going to provide them, and have a tracking mechanism to make sure that the techies don't deviate from that voice of the customer (Koch 2002 on-line).

In other words, they had treated the project as an exclusively technical issue and failed to understand the ways in which the Virtual Workplace needed to be integrated into the individual work practices of the suppliers. The Virtual Workspace Project is still alive, although for a time it looked as if its demise was certain, but it is moving more slowly, and is far less ambitious, than it was. It has begun to take collaboration seriously.

The Virtual Workplace Project is not the only B2B electronic exchange to fail because it was treated as an ahistorical, asocial, technological artifact. Fundshub, the B2B exchange for the financial industry, has had similar problems, as have many others. It seems that there is a fundamental problem in the structure and conceptualization of the exchanges that prevents them from operating as their developers expect and promise.

TEXTS, 'CYBER TEXTS' AND THE TEXTUAL PRACTICES OF COLLABORATIVE PROBLEM SOLVING

It is helpful to think about the Collaboration Manager as a more or less conventional workspace. Another way of thinking about it is, of course, as a 'cyber text', a more or less conventional written text. In some respects it is not a conventional written text at all. The parameters of the text, the way a person enters and engages with it, have been set by technicians responding to a range of pressures that have little to do with any specific project and may seem entirely arbitrary and not particularly useful to any individual group of participants. The text is represented as more like a conversation than a document; people 'take turns' and may be relatively conversational in the tone of their contributions. Sometimes contributions might be more like data—figures, specifications, diagrams. Sometimes they may be more like 'thinking aloud' half formed ideas. When the problem is 'solved' the text might seen to dissolve into the ether. Despite the private, conversational, 'work in progress' feel of the text it is by no means private. A range of people not nominated as team members may be visible or invisible observers; the full (written) record of exchanges belongs to the automotive company.

To engage with the Collaboration Manager in even the most minimal way, a participant must be able to read and to write. She will learn that she has been co-opted to the team through written text (probably E-mail), she will navigate her way to the electronic workspace with written directions, she will learn how to engage with the workspace partly through explicit directions and partly by becoming attuned to the written comments and replies her colleagues make and adapting her textual practice in response. If she is to make any contribution to the project at all, she must write. She will represent herself, her identity within the project, by writing (and perhaps a photograph), and she will 'meet' her colleagues and form judgments about them by literally 'reading' them. Problem solving with the Collaboration Manager is essentially a matter of collaborative writing. When the problem is solved the company has a permanent and complete electronic documentary record of the problem solving process.

The Collaboration Manager and the *Autoco Quality Manual* are in several important respects similar. Both represent attempts by global companies to regulate the work processes of remote trading partners by standardizing practice to that of the global company. Both are templates, primarily designed to be read and interpreted, and used, by people who have had little or no input into their development and who are unlikely to know the contexts in which they arise or the conditions under which they were developed. And they both rely entirely on the reception and production of written texts with highly specified features, whether on the page or on the screen. The work processes they describe and prescribe are incomplete and, even if they weren't, they apply to a particular interpretation of a remote context, not to the specifics of local sites like the Exmouth Plant or the Harbor Mill. Indeed, AFM, with a history within the textile industry rather than the automotive industry is likely to find the whole process even more bemusing than organizations located more centrally in the automotive industry. Even with the best will in the world, people like Bill and Matt have to 'animate' texts like the *Autoco Quality Manual* and the Collaboration Manager. They must modify and adapt the various work practices described in them to make an approximate 'fit' with local conditions, and they must 'fill in the blanks'—bridge the gaps in the work process about which the template is silent.

So far I've argued that the Virtual Workspace needs to be thought about as both a 'workspace' and as a kind of 'cyber text'. On the one hand, thinking about it as a workspace highlights the way in which it is shaped by the historical and material conditions which surround its production. On the other hand, thinking about it as a text highlights the absolute importance of established and new kinds of reading and writing in making connections between people in on-line environments. If the text is to have an impact on local workplaces then it must be brought to life by people like Bill and Matt and Sally and Margaret. Although the text is presented as a neutral and entirely rational way to solve problems, as we have seen it is neither neutral nor rational, and it deals only with those elements of problem solving that are amenable to computer programming. Some of its limitations are imposed by the computer program itself, and what the technicians can make it do from a technical point of view. Other limitations are more the result of pretending that the process of collaboration is a neutral process and that there are no tensions, compromises or negotiations to take place; it is as if the problem-solving process will lead inexorably to the single correct solution. What this means is that people on the ground have to find ways to make the Collaboration Manager work. To do so they must resolve fundamental tensions inherent in collaborating across time and space, in joining up the local and the global. I want to look at two tensions in particular, the tension between standardization and customization and the tension between erasing conflict in virtual workspaces and managing it.

STANDARDIZATION AND CUSTOMIZATION

The first aim of the Covisint Virtual Workspace Project is the standardization of products and processes across the automotive industry. With the production of the global car distributed around the world it is easy to see why a certain degree of standardization would be absolutely critical. But globalization is not all about standardization; it is also about customization. The production of the global car is outsourced to different locations around the world because different locations hold specific local advantages. If this were not the case then it would make more sense to reduce transport and communication costs and locate all operations in the one place. For instance, a labor-intensive operation may be located near a cheap, literate and relatively compliant labor force. A plant manufacturing heavy parts may be located close to cheap, efficient and reliable road, rail or sea transport. A technical design facility may be located near a concentration of related design companies and specialist university research departments for no better reason than proximity seems to generate certain kinds of creativity. Silicon Valley, the global hub of computer design and development, is a telling example of this kind of concentration. A Head Office may be located in a major commercial city to provide access to information networks, reliable broadband and other communication technologies, appropriate high status office space etc. A branch office, on the other hand, could be located in an industrial estate on the edge of a major city, within reasonable traveling time of affordable housing, schools, universities and hospitals, thus attracting a workforce of educated young family people. The point is, location matters. And, while all these operations need to be capable of communicating with each other, or there will be no car, they will also develop practices and processes of their own, that reflect the particular conditions of each location.

While standardization is the aim, however, it is only the aim up to a point. While the major car companies may seek the economies of standard parts and processes they do not wish, themselves, to standardize. Their market share relies on customization and differentiation. They need to produce cars designed for a particular market niche that are distinguishable from their competitors. Similarly, suppliers do not wish to compete with each other exclusively on price. They need to offer products and services that are clearly distinguishable on a range of measures, at the highest price they can attract.

So, while the Virtual Workspace aims to standardize products and processes, it can only ignore the pull towards customization, it cannot eradicate it. The tension between standardization and customization must be addressed by the virtual workgroup outside of the framework offered by the Collaboration Manager. They will need to develop their own practices and processes to deal with the specific tensions of their project. They will need to improvise.

THE PROBLEM OF FRICTION AND THE ISSUE OF CONFLICT

The second aim of the Covisint Virtual Workspace is the elimination of friction. The nominated cause of friction is the inadequate transfer of information. The argument seems to be that if all of the workgroup has direct access to the same information at the same time then friction will be eradicated. This is unlikely, and is probably also undesirable. Friction is not always a bad thing. While it can certainly be negative and destructive, especially when a team is working to deadlines, the 'creative abrasion of difference' can be productive, generating innovative solutions to intractable problems. The Virtual Workspace is a workspace socially constituted like any other, and conflict will feature in it, as it does any workplace. In trying to eliminate friction the Collaboration Manager ignores it, and, what's more, fails to provide a framework to manage conflict.

When a workgroup comes to interact in the Collaboration Manager, they meet in an environment which has been structured as if standardization must always occur at the expense of customization, that friction has been eradicated by the direct and timely communication of information, and that local workgroups and alliances are irrelevant to participation in the project. Since this is almost certainly not the case, individuals making up the workgroup must find ways to reconcile the tension between standardization and customization, manage and use conflict and integrate the work of the virtual workgroup with that of local workgroups who are producing component parts. While they have to do this in the Virtual Workspace, they cannot rely on the Virtual Workspace to do it. They must develop the skills themselves. Very often, because of the heavy reliance e-commerce places on written texts, these negotiations must be textually mediated and produced.

MANAGING CONFLICT AT LOCAL SITES

A significant feature of the Virtual Workspace is that, like the *Autoco Quality Manual*, it aims to shift legitimacy from the local site to the global site, and this has ramifications for many people at the local site, not only those who actually engage with the Virtual Workspace. It is helpful to look at the ways in which a local workgroup collaborated to integrate the *Autoco Quality Manual* into local practice, managing the tension between standardization and customization and bringing a range of sensibilities and strategies to the management of conflict.

During the Action Learning Team Meeting discussed in Chapter Four it seemed clear that, while Sally and Margaret made explicit efforts to shift legitimacy (that is, whose knowledge counted and how it should be framed) from the local site (Bill) to the remote site (Autoco), Bill was not prepared to accept the shift. He rejected the idea of a 'team' collectively producing knowledge and asserted 'experience' as the basis for knowledge production. He was prepared to have his knowledge written down, but he was not prepared to be the one who wrote it. He delegated this job to Margaret, and Margaret accepted this role, 'gradually translating it on to here'. In this case Margaret is acting explicitly as mediator, taking local knowledge expressed in local work practices and discourses, and 'set[ting] it down' in ways that are acceptable to Autoco. Here all the participants in the team meeting collude to 'work' the available discourses and produce knowledge which has legitimacy at Autoco without completely ceding to Autoco the right to be the agent of legitimization. Bill retained the authority to determine 'what counts' as knowledge in his own domain of expertise at AFM, but that authority, having once been questioned, is now permanently under threat. As Matt reflects below, writing is at the crux of the conflict for Bill, and it is up to the rest of the workgroup to find a way to make Bill's knowledge available:

> If you notice, at the meetings, as soon as we talk about writing things down, Margaret does it. Now . . . I don't object to that. But it defeats the purpose of what we are all here for. So, Bill's the sort of guy we've got to bring into the discussion by picking his brain. . . . Pick that, and Baz's another one, pick their brain and get it out of them and that's the site where we are going to do it. . . . Bill's very quiet. I think he's intimidated. And he has, you know, literacy skills that he thinks are not up to scratch and when he . . . thought they would be on display

he pulled out. The pressure of the buggies project, you see, you know, he's obviously feeling the pressure from doing that.

Matt recognizes that the Action Learning Team meeting is a site of knowledge production, and a complex one. Writing, the documentation of knowledge, is a critical element in knowledge production because it makes local knowledge available to the team and the company (and, ultimately Autoco). Matt identifies Bill's apparent unwillingness to write down what he knows as a problem, 'it defeats the purpose of what we are here for'. It is not just writing that is the problem, however. Matt attributes Bill's silence in the discussion in the meeting to his sensitivity about his writing skills, 'when he thought they would be on display he pulled out', so the meeting can't get at Bill's knowledge either through writing or through talking in the normal way. Matt proposes that he and the other members of the group have the (discursive) task of 'picking his brain', and that the team meeting provides the obvious window of opportunity, 'that's the site where we are going to do it'. The Buggies Project is, in Matt's view, clearly putting pressure on Bill. A process that was intended to make the knowledge of individual members of the group more easily available to the whole group is in practice exerting the opposite pressure; the group is denied the expertise that Bill would formerly have made available as the recognized expert.

It seems that Matt is correct in assuming that the Buggies Project is putting pressure on Bill:

> I had them here with the meeting on Friday to discuss these buggies what I'm after and they're going way above my head now. I don't even know what they're talking about. I mean, I worked out all the pricing, what I lost and what I think we're losing, what the company's losing, and they want to work it into hours, not the price dollars. I said 'What are you talking about? The bosses understand dollars not just hours. . . . So, I don't know what I can do now but I've been losing a little bit of sleep over it and I've never done that before. Why should I lose sleep at my day and age, worrying about the job?

Like Matt, Bill attributes pressure from the Buggies Project to his own lack of confidence in the new discursive practice of the team meetings. Bill's concern is not with the need to write, however, but with the words themselves, the words in which he must frame his knowledge if it is to be recognized as legitimate by his colleagues on the team. Although the Buggies Project is centrally in his area, and although he has been working in the area for 30 years, and he is acknowledged as the expert by people like Matt, he says 'I don't even know what they are talking about'. Bill's concern is that, despite his expert knowledge ('I worked out all the pricing'), he is silenced at team meetings because he can no longer understand what is going on or express himself in ways that are recognized as legitimate in the *Autoco Quality Manual*, although the team knows what he is talking about and is aware of the value of his knowledge. Despite this, he asserts that 'the bosses understand dollars, not just hours'. He is concerned that, because he no longer knows what the right words are, his expertise and his judgment about what 'the bosses understand' is no longer accepted.

Margaret is the workplace educator charged with the responsibility of inducting Bill and other workers into problem-solving teams. She, too, is worried about the 'Buggies Project':

> It's part of my worry at the moment, how things are going with Eight Step [problem solving] Plans. I'm too much involved at the moment . . . The fact is that we don't have formal meetings but it's more that I'm the link between various members and sometimes we get together informally and they talk about things and I write things down. . . . Often it will be that I will be talking to a couple of the warpers and then Bill will come up and then we'll say 'well, about that problem we've been working on' and then start to talk about it. And I always have my notebook with me so I'm writing things down about it. But I think then I'm seen as probably guiding and steering it. Particularly Bill, who'll say 'We've got to do this and Margaret, have you followed up?' on something or other. So in a way it's hard then to say 'Well, this is your team and you have to be doing this'.

As a workplace language and literacy teacher Margaret's job is to teach the textual practices required in team-based problem solving, especially the kind of writing demanded by the *Autoco Quality Manual*. From her perspective, however, the problem is not that Bill cannot write what is required, or even that he is embarrassed about his writing. The problem is that he is reluctant to relinquish his traditional working identity, as the source of knowledge and final authority about warping, to take up the new 'working identity' offered to him in the problem-solving team; he doesn't want to be a Team Leader. In rejecting the writing demands of the new problem-solving processes, Margaret argues, Bill is rejecting the role of Team Leader and with it the implied requirements to define a new, less obviously hierarchical working identity, and to renegotiate new relationships with his colleagues in the team. Knowledge is legitimated in a web of formal and informal narratives in local workplaces, it is not fixed, it is always being contested and negotiated. There is no reason to suppose that it will be different in on-line environments like the Virtual Workplace.

The Action Learning Team meeting is a local site in which Margaret and Sally, as workplace educators, are positioned as 'discourse technologists', actively intervening in the discursive practices of work at AFM to shift the policing and legitimating of knowledge production from the local to the remote site. This is not as simple a matter as it first might appear to be, and it is by no means clear that Sally and Margaret have succeeded in this aim, or that they or the other members of the team are in fact willing to cede authority over 'what counts as knowledge' to Autoco. While they recognize the imperative of producing the texts that Autoco demands, all the members of the team nonetheless collude to find ways to preserve Bill's authority in the local workplace, at least for the time being. If we think about knowledge as Smith does, as always produced locally in textually mediated social action, we can see why it is that, while it is a relatively simple matter to shift the policing of the discursive practice of knowledge to global authorities, it is no easy matter to shift legitimacy. Nonetheless, workplace educators occupy a potentially powerful position in workplaces like AFM. When they mediate local and global discourses they play an important part in shaping 'what counts' as knowledge in a Knowledge Economy, and who can say so.

New work practices are generating new textual practices, and these new textual practices signal risks, for people like Bill, and opportunities, for people like Matt. At least part of what is at stake here are working identities and working relationships as they are produced in the intensely pedagogic space that the contemporary workspace has become. The next chapter looks more closely at how working identities are taken up and put down in the pedagogic space of the Action Learning Team. Traditional pedagogic working identities, like those of Margaret and Sally, are also under pressure. In Chapter Seven I look at the ways workplace educators engage with the texts of the global workplace.

CHAPTER SIX

Learning how to be
the textual production of working identities

To put it simply, learning is the new form of labor.

(ZUBOFF)

INTRODUCTION

When Bill resists the role of 'team leader' of the Eight Step Problem Solving Team (in the last chapter) he is not doing so simply because of the reading and (more particularly) writing demands that are made of him, demands that at least some of his colleagues believe he cannot meet. I observed Bill on The Floor for over eight months, and it seemed to me that he was well able to read and write; certainly he was literate enough to deal with the routine conditions in which he worked. It was not reading and writing *per se* that was a problem for Bill, it was the uses to which these textual practices were to be put that presented a constellation of problems.

The first problem was that Bill was asked to write his knowledge down. This is not as simple a matter as it sometimes appears to be. Professional technical writers find it a complex task to render highly contextual work practices in written text. It involves such a fine reading of the context, and such an explicit account of the resources people use and the decisions they make of which they are only partly conscious, that it is customary to note how incomplete textual accounts of work practice are. This is not Bill's professional training, and it is not surprising that he balked at the task. Bill's knowledge did not sit in his head as paragraphs of text, or even as dot points. There was a lot of translating to do if even a part of his knowledge was to make it to the page.

The second, and more critical, problem was that Bill did not value written knowledge above embodied or contextualized knowledge in the way that Sally and Margaret, or even Matt, did. While Sally and the others might believe that writing down knowledge is a way of making it visible, raising it to the status of legitimate learning, for Bill the opposite seemed to be the case. Like many others on The Floor (even the young people from the Harbor Mill), Bill commented repeatedly that written texts could not capture his knowledge, or any other working knowledge worth having. From his point of view texts like these trivialized his knowledge, suggesting that someone could, simply by reading directions in a manual, accomplish the creative knowledge work that he did on The Floor.

The third, and most significant, problem concerns the implications of the primacy of written knowledge for Bill's authority and identity in the workplace. Bill well understood that the dominance of the written text was fundamentally about restructuring authority in the workplace and shifting legitimacy from the local site to the remote Head Offices of companies which did not even employ him. He understood, too, that this shift was a demand of the automotive industry and that the rules and processes of the automotive industry were being inserted into his workplace, previously understood as part of the textile industry. Bill was proud of his work and of his status as a Machine Lacemaker within the textile industry—a role that did not exist in the automotive industry. Accepting the dominance of written knowledge seemed to him to compromise that hard-won status. It is not surprising that Bill resisted the identity of team leader (in the sense of facilitator, or first amongst equals) and hung on tightly to the role of expert knower.

In the face of the kind of trenchant opposition that people like Bill offer, there must be good reasons why global corporations, industries and governments support large-scale industry training programs that aim to induct people into the textual practices that appear to constitute so much of contemporary work practice. One of the most critical reasons, as I have discussed in previous chapters, is that the successful functioning of globally distributed supply chains depends on ICTs, and ICTs rely on reading and writing for both the standardization of practice and for improvisation and problem solving. Since global corporations have outsourced so many of their operations, they must use reading and writing to achieve the levels of knowledge production, surveillance and regulation they require. If knowledge and practice are to be made both explicit and abstract, then people have to be able to read and write in standardized ways.

But, teaching people to read and write is not, as Bill is well aware, a simple matter of adding to their repertoire of textual practices. The texts Bill is invited to engage with are not neutral, nor are they intended to be. New ways of reading and writing entail new social relationships at work they foreground new roles and new subject positions and disrupt established roles, relationships and hierarchies of knowledge and the power and influence that go with them. They entail a fundamental shift in legitimacy from the local site (like the Exmouth Plant) to the remote site (like the Head Office of Autoco), and they imply fundamental changes in the ways people like Bill and Mary and Baz and Grace occupy their local workplaces and relate with each other within them.

One of the strategies used to make this kind of the shift happen is to associate formal workplace training with reading and writing. As I discussed in Chapter Three, traditional, formal and informal workplace education has happened 'on the job', or very near it. People learned from an apprentice master, or from 'sitting next to Nellie'. Reading and writing were relatively minor aspects of most workplace training programs. Workplace education programs associated with Quality Management, drawing on concepts like 'the Learning Organization' and 'Lifelong Learning', are about transforming knowledge through reading, writing and new ways of talking. The focus of these programs is on decontextualized, generic and transportable knowledge that is embedded in textual practice, not on knowledge embedded in local workplaces (or even in specific, transient, distributed supply chains). Within this context, workplace education is not intended to teach people to do new things, or to do old things better; it is to teach people to do things with words. This is a fundamental move that is centrally about constructing new discourses and building new identities and new relationships at work.

This chapter is about the ways that learning and knowledge are conceptualized in the Action Learning Team meetings, about how new roles and identities are made available, and how different people, working by themselves and together, appropriate and adapt them to their own contexts and purposes. The role of workplace educators in the process is obviously important and, while I refer to them here, I will attend to the challenges they face in more detail in Chapter Seven.

To begin with, I want to look at the different ways in which the move to 'paperwork' is understood, by proponents of various forms of structured 'Quality Management' and by the workers whose task it is to enact the Quality Management texts. Then I will look at the very first meeting of the Action Learning Team, where the task of shifting legitimacy from the local to the remote site is unambiguously presented. I move then to look at the ways in which four workers, participants in the Action Learning Team and located at the Harbor Mill, appropriate the language and processes of Action Learning to try to find a way to integrate the demands of the global workspace with the practices and processes of work on the factory floor, developing new textual practices, and new identities, in the process. This involves these participants in improvising new textual practices, not in order to communicate with workers at remote sites, but in order to communicate with each other and the members of their own work teams. The discussion they have with Sally, when they are physically located in their own workplace and using their own words, lays bare the contradictions of the model they have been offered. Even so, Ben, James, Peter and Grace are inventive in mobilizing the resources they have at hand to try and make the Eight Step Problem Solving Plans work.

PAPERWORK AND KNOWLEDGE WORK

Manufacturing workers are no strangers to paperwork. Continuous Improvement regimes, in place in varying degrees for decades, demand that people document what they do, what they document, and standardize the documentation across workplaces, industries and global supply chains. In the textile industry for instance, Folinsbee (2004) describes what is required by Statistical Process Control (SPC), a requirement for the mandatory ISO (International Organization for Standards) certification in a number of contexts. SPC requires workers to:

> record production data. In the weave room it is the width of the product being woven and the number of picks per inch. In the warp room it is the tension of the warped yarn across the width of the material... (: 70).

Operators are given perfunctory training in the details of this kind of data collection—'Everyone in the plant gets four hours of SPC training' (: 70)—but they are not told why they are collecting the data, or what particular use is to be made of it. They generate the data, but they don't use it.

From an organizational point of view, the critical thing about data collection is that it is essential to satisfy certification requirements; without certification it is almost impossible to survive in a global market. Organizations are required to:

> specify, implement, monitor and record their compliance with Standard Operating Procedures in all areas of the work process. Compliance with all these steps is enforced through an on-site inspection called an 'audit', leading to official certification by various national or international bodies like ISO (Jackson 2004:9).

For many workers this kind of record keeping seems trivial, and a pointless waste of time. They do not decide which data to collect, they are rarely encouraged to make use of the data to inform their own problem solving, and they are suspicious that the main point is surveillance; they must collect the data in order that their supervisors can check that they are doing their jobs. It is not surprising that some workers take pride in subverting the process, by, for instance, completing process checklists well after an order has been filled.

Given the resistance to these kinds of documentation demands, it might be surprising to learn that many people initiate their own, private, complex, systems of written documentation at work, and have done so for many years. One example is Mary, the Mending Room Supervisor at the Harbor Mill. When she agreed to teach me to mend, she took her Mending Manual from under her worktable and gave it to me to look at. There are no formal manuals for mending, Mary had prepared her own, continuing a tradition begun by Mary's own Mending Room Supervisor forty years before. Mary had placed samples of fabrics exhibiting different faults inside plastic sleeves so the menders could see the fault, as well as 'the mend', take the fabric out and examine the back, and check it against the fabrics they were working on. Next to the faults she had written explanations of the possible techniques that might be used to mend them, and how a mender might decide which to use. The Manual included an additional, extensive, list of faults, possible causes and 'mends'. Mary updated the Manual regularly to provide even the most experienced menders with the information they needed to deal with newly developed fabrics and designs. Once, several years before, Mary had shown the Manual to the Production Supervisor to demonstrate the kind of knowledge people needed to mend, but he had laughed at it. Mary and the other menders were offended and kept the Manual to themselves after that.

The Mending Manual is not unique; Folinsbee (2004) reports that workers at Texco keep their own unofficial spiral notebooks with their own (non-standard) operating procedures, calculations and sketches which record and perpetuate local practice. These, too, are private collections of knowledge which may be shared amongst friends and workgroups but are not made available to outsiders. Even in high-tech industries like PaceSetters, workers who routinely use E-mail, SMS, and palm pilots, keep spiral-bound notebooks in the glove boxes of their cars to record the information that really matters to them (Farrell and Holkner 2004). None of these texts provides a comprehensive account of work practice, and they can be difficult to understand outside the workplaces in which they are created. They are all incorporated into local work practices and processes and represent a critical part of local working knowledge.

Writing about work in this way is a private activity; it arises in specific local work contexts because people need it, and people decide for themselves how and when they will integrate it into their own work practices. Generally, this kind of writing is confined to individuals or to specific work groups. The individuals are completely in charge of the texts, deciding for themselves what counts as useful knowledge, how it should be recorded, and who should have access to it. While texts like these are generally invisible to management (and are often seen as a guilty secret by workers), they are in direct competition with Standard Operating Procedures and other generic forms of process documentation demanded by Continuous Improvement systems.

Mary, unlike some other workers at AFM, is scrupulous about filling in the formal documentation for ISO accreditation, but, like many workers at AFM and elsewhere, she doesn't trust it. As Supervisor of the Mending Room one of her tasks is to keep track of the rolls of fabric. She has two methods for doing this. First, she enters information about each roll in the computer database, so that workers in other departments, and in other plants, can regulate their work and locate specific orders. Second, she keeps a more complete, handwritten record of the mending process, recording the time a roll arrives in the mending room, and the time it leaves, the condition of the fabric, the name of the mender assigned to it and the time taken to mend it. This second account is for her own purposes, to check the accuracy of the data base records, to provide information that will regulate the work of the mending room, and to provide the information she needs to defend the menders against accusations of slow or sloppy work. When a roll of fabric is returned, this record ensures that it is generally returned to the original mender, to complete her work.

People read and write at work, and they use their own idiosyncratic textual practices to help them to solve problems, to store knowledge, and to improvise when they need to. For Mary, as for many other workers in high- and low-tech industries, 'paperwork' is part and parcel of local work practice and reading and writing are part and parcel of local knowledge production. In other words, reading and writing are integrated into local work practice as part of 'knowledge work'. The kind of reading, writing and talking associated with Quality Management, the kind that involves problem solving and knowledge-building protocols (eg Eight Step Problem Solving Plans, Problem Solving Plus) is, however, another kind of textual practice altogether. These protocols require people to engage in externally imposed textual practices which involve reconfiguring workplace and workspace relationships.

THE TEXTUAL PRACTICE OF 'QUALITY'

'Quality Management' is generally recognized as an umbrella term that covers a range of business management systems popular over the past fifteen to twenty years or so. For my purposes here it covers a raft of movements including Just In Time, Kaizen, Quality Circles, World's Best Practice, Reengineering, Learning Teams, and many movements that are more transitory, and more local, than these. While there are some differences, these movements share a number of features:

* Explicit identification and measurement of customer requirements
* Creation of supplier partnerships
* Use of cross-functional teams to identify and solve quality problems
* Use of scientific methods to monitor performance and to identify points of high leverage for performance improvement
* Use of process management heuristics to enhance team effectiveness
 (Hackman and Wageman, as cited in Bruce Wilson 1997: 78)

Quality Management regimes and rhetoric are pervasive. They appear in management textbooks, political speeches on all sides of politics, some industrial union documents and many education and training policy documents. These are the narratives that dominate 'fast capitalist' texts, composed of 'part "reality", part prediction and part hope' (Lankshear et al 1997: 84). A feature of many Quality regimes is that they come with a moral imperative; they promote shifts in working identity and working relationships as being unambiguously good for the worker (as a worker and as a human being) as well as being unambiguously good for the organization. Elsewhere I have referred to Quality Management as a modern version of the medieval 'morality play' in which 'Everyworker' undertakes a journey from darkness to enlightenment (Farrell 1999).

Quality Management is fundamentally about providing new identities for workers and ensuring that they are, to a significant extent, self-regulating identities. Hammer is one popular writer in this field whose books promoting the concept of business 're-engineering' continue to be influential. He provides a seductive account of the radical shift in identity required in the Knowledge Economy. He explains that there is no longer a place for the traditional worker, who focuses on product, in the 'reengineered' (or 'high performance', or 'fast capitalist') workplace. The new focus is on constantly evolving processes, and this requires a new kind of worker:

process centering is eliminating both the traditional industrial job and with it the very concept of the industrial worker (: 33).

In Hammer's idealization of the contemporary workplace, embodied knowledge takes second place to symbolic, encoded, or abstract knowledge; all workers work 'with their heads, not just their hands'. He makes a stark distinction between the old kind of industrial worker, whom he characterizes as less than human; 'a kind of organic robot, operated by a manager via remote control' and the new kind of autonomous, educated worker who is: 'a professional worker. . .an independent human being' (: 46). For Hammer it seems that it is only in high performance workplaces that 'professional workers' are able to be fully human.

The construction of new knowledges is intimately associated with the construction of new identities, and the role of education (as opposed to the traditional vocational 'training') is central. Lifelong learning is the critical commitment which will keep a professional worker in a state of constant transformation. A worker is merely trained, Hammer argues, but a professional is always learning and always making new knowledge.

For the worker, the promised benefits of embracing this new work order are profound. 'Everyworker' will be removed to a new spiritual level in which personal and working identities are fused:

> Professional work is not an activity performed a certain number of hours a day, but one's persona, one's essence (: 49).

Managers will voluntarily relinquish authority. They will coach and design rather than organize and control.

These new workplaces will un-tether themselves from their historical and social foundations to create a society that is, finally, fair and just:

> Connections, background, ethnicity, race religion and gender no longer count. The process organisation is a true meritocracy, the original American ideal and the realisation of Dr Martin Luther King's dream that men and women may be judged only by the content of their character (: 264).

People who resist this new world order are, it is implied, not only stupid, they are bad.

For the workers at AFM, these Quality discourses are embedded in the Autoco Quality document and the Action Learning Team program.

PLACE, SPACE AND THE ACTION LEARNING TEXT

When the Action Learning Team met formally it met in luxurious surroundings. The Mansions is a stately home converted into a conference center, located half way along the 100 kilometre stretch of highway that separates the Exmouth Plant from the Harbor Mill. The conference room (which is often used for wedding receptions on the weekends) is thickly carpeted, heavily draped, and furnished with gilt-trimmed chairs and tables. White boards and video-players were wheeled into the room for Action Learning Team meetings, and there was always a good supply of butchers paper and marking pens. From the windows participants could easily see beautifully kept gardens and the occasional ostrich or zebra strolling around the free-range zoo associated with the conference center.

The choice of venue was a deliberate one, made by Sally and Michael to provide what they believed to be an environment conducive to the right kind of learning. Both the Harbor Mill and the Exmouth Plant had meeting rooms large enough to hold the group, and the team could have alternated at each site to share the burden of the journey, but Sally and Michael were eager to remove the team from their immediate physical environments and from the 'cultures' of the two workplaces. Like the designers of the Virtual Workplace, Covisint, part of Sally and Michael's purpose seemed to be to create a neutral knowledge-building space, a space where outdated practices and processes would be erased and friction (between individuals and between the two workplace cultures) would be eradicated.

All the same, many of the participants were reluctant to leave their workplaces and were acutely aware that at the best of times they were talking and writing about their work rather than doing it, and the build up of work would be waiting for them when they returned to the floor the next day:

> Matt: I can't be doing something else if I'm here. And if they come along tomorrow and say 'why isn't the recovery plan done?' I'll say well, I just had to sit up here for three hours yesterday . . .

The idea of the 'Action Learning' Program was introduced by an external facilitator in the first session:

> What this group is about is getting you all off the floor; for you all to get this plan of action, to come together to reflect and understand, to figure out the whys and wherefores. So it won't be about somebody like me coming in and saying 'this is

what you have got to learn, take notes.' That is not the way of the worker. *You have got the knowledge to figure out what you have to do, to figure out what you have to learn to find out, and you can figure out your own ways of learning* . . .

This introduction suggests that the participants will be involved in learning based on 'action', learning that values and calls on their experience and knowledge, rather than abstract school-like knowledge. In fact, the very first activity required the group to 'take notes'. Sally invited people to:

reorganise yourselves, shift the chairs and tables if you want to, mix up as much as you can so that you have got a mixture of males and females in the group, managers and underlings in the group, Exmouth Plant and Harbor Mill in the group, and just come up with [write on butchers paper] a brief, clear definition of what action learning is.

In this single initial task established relationships and longstanding alliances were disrupted, and text-based, abstract learning, and encoded knowledge, was established as the kind that was going to count. As Sally walked around the room she noticed that people were commenting on the practical implications of action learning at AFM:

Can I just break in here? Instead of getting too specific at this stage about what the implications are for AFM can you just come up with a very clear set of guidelines or explanations about what Action Learning is, and why it is different to anything else. So, if someone at work wants to know, 'Well, how is this different to what we've been doing in all these other training sessions . . .'

Righto. Let's see what everybody thinks. You might like to just come along and have a read of it. It's small but you will get some idea and I think, have a look at what your group said. Then go back, and in your own notebook, or on the cover of your file, or something, just quickly write down a one sentence definition for nobody else but you.

Bill: I'm going to cheat.

The Mansions was chosen as a neutral environment; the facilitator asserted that 'you are all equal here'. While it was a comfortable break from routine for many of the managers it was, nonetheless, an exquisitely uncomfortable environment for many people who worked in the weaving and warping sheds. In each contrived group, it was a manager who scribed the definition of 'action learning' and came to pin it on the wall.

While most participants were sceptical of this eventuality, they did finish the first session clear in their understanding that action learning was about writing, and that writing was a way of standardizing knowledge and practice. Generally speaking, people from the sheds saw this as a means of generating local knowledge and regulating the practices of their own teams, perhaps formalizing the private writing that was already a part of work practice:

> James: Maybe with Action Learning it might involve more people so they're learning as they're doing their own job and passing on that knowledge too, hopefully, where it becomes written form so it becomes a standard, and so forth.

Managers, on the other hand, saw it as a way of satisfying the requirements of senior management and global quality frameworks:

> Michael: Yes, one of the big attractions for me is that it fits straight into the QOS cycle . . . and the other thing I like about it is that it is doing things which are immensely attractive to senior management.

The Mansions did generate some unexpected other conversations. Michael, the HR manager, spent a whole lunchtime talking to Andrew, the new warper, about dreams, and reading his poetry.

NEGOTIATING IDENTITY IN GLOBAL WORKSPACES

Bill is not alone at AFM in resisting the identity of Team Leader as facilitator, and the constructions of knowledge, legitimacy and power that go with it. The young people at the Harbor Mill also struggle with this concept, focusing especially on the way in which the concept might sit next to their other roles.

The Autoco representative came to The Mansions to promote the need for the Autoco Quality Manual requirements to be integrated into the Action Learning Team meetings. Workers at the Harbor Mill understood the necessity for this, but they asked Sally to convene a meeting on the Floor at the Harbor Mill. They had concerns they wished to discuss. I've argued earlier that the Autoco Quality Manual is indicative of the texts associated with Quality discourses that emerge at AFM and that it is centrally concerned with discipline and control and only equivocally concerned with the production of the 'autonomous professional worker'. In presenting the following transcript segment I want to make the argument that the Team Leaders at the Harbor Mill take the Quality discourses seriously but that they also look to other workplace discourses, and other personal identities, to provide discursive resources that will lead to the production of new working identities around the title of 'team leader'. They use the cross-functional team meetings required by Autoco to collaboratively construct, not a single shared set of practices around 'team leader' but several possible ways of doing 'team leadership'.

THE TEAM MEETING

The conversation that follows is a record of part of a team meeting facilitated by Sally. Ben and Grace had discussed problems they had experienced with their Eight Step (problem solving) teams and had resolved to bring these problems up at the meeting. The essential problem, identified here by Ben and confirmed by Grace, is that they are positioned by local workplace discourses in one set of social and power relations, and by global Quality discourses (specifically the Autoco Quality Manual) in another, contradictory set of social and power relations. They are seeking ways of resolving the dilemmas that have arisen as they have attempted to simultaneously occupy the roles of supervisor and team leader/facilitator. The transcript records a critical moment in which the team open for scrutiny the usually implicit dynamic of identity formation, exposing the ways in which their established identities 'work' the new Quality discourse (instanced by the establishment of Eight Step teams) to fashion workable identities and workable social and power relations. Ben, Grace, James, Peter and Sally speak.

Ben, in his late twenties, is a Weaving Shed Supervisor who has worked in a range of textile manufacturing plants. His task is to supervise a group of weavers and to ensure that the production schedule is met. He is a qualified Textile Mechanic.

Grace is in her late twenties. She completed her training as a Chef before coming to work at AFM, where her mother had worked and her aunts and cousins are still employed. She learned the machines quickly and rose rapidly to the position of Warping Room Supervisor.

James is in his mid twenties and works in the small laboratory. He is a qualified Laboratory Technician. He is an active member of his local Christian church community and takes a leadership role in the church youth movement.

Peter is in his mid thirties and is Supervisor of the Finishing Department. He is responsible for ensuring that deadlines are met and machines do not remain idle. He has no formal vocational training.

1		Ben	we thought you know maybe maybe I should be the facilitator for
2			Grace's group or something where I'm away from the people a bit
3			and um
4		Sally	Yeah
5		Ben	just have a background in what's going on
6			but just sort of keep them on the right track and let them
7			they've got to really then rely on each other instead of relying on
8			the supervisor to do the work
9		Grace	Well I think kind of in the groups that are gonna come along

10		that's what's gonna have to happen. I mean I know the first ones
11		that start off, I think we have to go down this path to try to direct
12		people onto the path, and therefore we kind of will be in charge
13		of the meeting but then we have to get people to start their own
14		teams and us sort of just being a facilitator rather than
15	James	the team leader
16	[...]	yeah
17	Grace	I mean it's hard to get started I think that's where people are
18		having trouble and that's why they're kind of looking to you Ben
19		and you know things like that
20	Peter	I'm not the only one I'm having trouble maintaining the thing
21	[...]	Yeah
22	Peter	I just can't maintain it at the moment you know a couple of days
23		you know a couple of days crook there and you know just the
24		amount of work that builds up it just goes to the back of the
25		queue sort of thing it's shocking
26	James	so what you really want is the um you've got a a group you start
27		a group and you want one of those people to sort of come out
28		and [...] facilitate the group
29	Peter	just to maintain the group you know like just keep it just keep
30		the work flowing
31	Ben	What I'm trying to get across
32	Peter	'cause
33	Ben	is I'm too close to those people because I already go outside of
34		the group and then I'm their supervisor outside on the on the
35		floor where maybe if I was facilitating another group where I'm
36		not I'm not above them you know I'm not their supervisor or
37		whatever
38	[...]	yeah
39	Ben	um I can go back to my job they can go back to theirs and they
40		still um you know it's this their more their team than
41	Sally	yours
42	Ben	then everybody talking to the supervisor sort of thing you know
43	Sally	hmm
44	(James)	well in this Action Learning thing we we like like these these
45		problem solving things are supposed to be part of it alright
46	Sally	Yeah
47	James	so and what we're learning is how to learn this Action Learning
48		like how to apply it to ourselves
49	Sally	yeah
50	James	and also to facilitate it to other people
51	Peter	excuse me [ironically]
52	James	so what we're saying is in our groups we want the people
53		somehow to be [like that] you know like to learn actively like
54	Sally	yeah
55	James	you know like sort of and that's what we're trying to promote
56	Sally	yeah

57	James	and that's what we're trying to do
58	Sally	we've got to work out ourselves what it is you need them to do
59		and then
60	James	and get them to do that
61	Grace	yeah
62	Sally	yeah
63	Grace	and that's not easy
64	Sally	no not at all
65	James	I just thought that was worth identifying that was what we were
66		trying to do instead of saying
67	Sally	yeah
68	James	instead of saying we're trying to get them to take it over or
69		trying to get them to
70	Sally	yeah
71	James	be a bit more responsible we're trying to get them to sort of
72		change their attitude towards what they do

'Working' discourses, working identities

I have argued above that the Autoco Quality Manual positions workers in multiple and contradictory ways, and that it is this which creates a space in which AFM workers can 'work' discourses to create working identities with which they can live. Now I want to look closely at the moment in an Action Learning Team meeting where Ben, Grace, Peter and James explicitly 'work' the discourses that are available to them to try out 'alternate selves'. While there are traces of innumerable discourses washing through this site, here I want to attend to three that are obvious: the local workplace discourse, the global Quality discourse (including discourses around 'learning organizations') and a progressive adult education discourse that is closely associated with contemporary workplace literacy education. These discourses have important points of connection. They are each centrally concerned with values around knowledge and authority, they are centrally concerned with the social relationships which mediate these values in the workplace, and they each have embedded in them certain work practices and working identities. The participants in this Action Learning Team meeting call on each of these discourses as they work collaboratively, through the structure of the Action Learning Team meeting, to create new, transitory, working identities. Although, in this transcript they deal explicitly with the central problem of constructing new working identities to accommodate these new times, they seem to be heading for somewhat different accommodations.

(i) Sally: naming the job

Sally names her own job in relation to the Action Learning Team, and she names the job that the team members hold in relation to their individual Eight Step Plan teams. She labels these jobs 'facilitator', effectively conflating the work she does with the work Ben, Grace, Peter and James do with their Eight Step Plan teams (although in line 58 she seems to be ambivalent about this, moving from 'we' to 'you' as she articulates the aim of the Eight Step Plan teams). The Eight Step Plan teams were originally convened by 'team leaders', a job title that is firmly established in Quality discourses and more recently established in the local workplace discourse. Early in the series of Action Learning Team meetings Sally began to use the term 'facilitator' when she was talking about the position of team leader. She also used the term when she referred to her own role in relation to the Action Learning Team. While she never corrected anyone when they referred to themselves as a 'team leader' she always referred to them as 'facilitators' herself. By the time the discussion recorded here took place the term 'facilitator' had been comprehensively adopted by everyone in the Action Learning Team, so comprehensively, in fact, that it is used in opposition to 'team leader' (line 15).

The 'facilitator' is central to the workplace literacy education discourse in which Sally is most securely located. Knowledge and authority are central but problematic values in this discourse. While 'knowledge' is a core value, workplace literacy educators do not, and cannot, have detailed understanding of the knowledge, skills and work practices in all the places they teach. The term 'facilitator' is often preferred over more traditional labels like teacher or educator because it gives a name to the kind of knowledge the workplace literacy educator does have-knowledge about process and knowledge about how to assist others to articulate their own tacit knowledge. The term 'facilitator' also obscures the authority of the position; while it echoes the relationship between teacher and learner of established education discourses it constructs that relationship as an emancipatory one in which the facilitator/teacher liberates the worker/learner by making explicit their tacit knowledge and providing tools for them to make their own decisions. These identities, and the social relationships that underpin them, seem on the surface to sit comfortably within idealized Quality discourses that value knowledge, identify the worker as learner, and encourage the development of flat organizational structures in which the 'professional worker' operates independently.

(ii) James: learning the job

James has taken up the 'learner' identity made available in the workplace education discourse offered by Sally. What he is working through in this excerpt is what it is that the learner is expected to learn. As he works this through, with

Sally's encouragement, he comes to the conclusion that 'we' are trying to change attitudes, their own and those of their team members. Between lines 44 and 72 James articulates the values and orientations associated with these discourses, with Sally interjecting with support as a 'coach' might. He identifies the role of 'facilitator' as centrally about learning, names the practices associated with it as 'facilitating' and is the only participant to name their group activity by its formal name, 'Action Learning'.

'What we're learning is how to learn/this Action Learning/like how to apply it to ourselves' (lines 47,48).

Later (lines 68–72) he plays with the name in an attempt to generate the appropriate verb to describe what the aim of the teams is, saying that they want people to 'actively learn'. However, while James embraces the learning organization and the values and orientations that go with it, he nonetheless treats the 'Action Learning' in much the same way as 'structured problem solving' is treated in the Autoco Quality Manual rating scale. He refers to 'Action Learning' as something, perhaps a set of cognitive activities, that can be 'applied' to themselves and to the people in their groups. James struggles, and fails to find words for what this might mean for the social relationships and practices in the workplace. He says that they all want the people in their groups to assume worker/learner identities although he cannot say how this will happen, they are 'somehow to be', and, pressing himself to be more specific, he tries again with a new configuration of 'Action Learning' to 'learn actively like'. In line 55, encouraged by Sally, he tries out 'promote' to name their task in advancing Action Learning in the Eight Step Problem Solving teams, but this doesn't seem right either. James is trying to call into being the kind of relationship between 'facilitator' and team that would bring about a change in the values and identities of the team members, that would induct them into the 'Action Learning' discourse, that would make them willing learner/workers. Paradoxically, this will require a level of leadership: 'Sally: "we've got to work out ourselves what it is you need them to do and then"', and a degree of coercion: 'James: "and get them to do that"' that does not sit easily with the social relationships between autonomous, professional, learner/workers that are implied in the Quality management discourse. At the end of this segment James has explicitly rejected 'autonomy', recognising, perhaps, the veiled challenge to traditional lines of authority that Ben and Grace have articulated. He acknowledges that they (as facilitators) are invoking the power some of them (but not James) have, as supervisors in the local workplace relationship between supervisors and workers, to unobtrusively shape the working identities of their subordinates:

James	instead of saying we're trying to get them to take it over or trying to get them to
Sally	yeah
James	be a bit more responsible we're trying to get them to sort of change their attitude towards what they do (lines 68-72)

(iii) Peter: doing the job

Peter, on the other hand, does not express any conflict between his role as facilitator and his role as Supervisor, in fact he talks about facilitating the group as one more item in the heavy load of responsibilities he carries as Supervisor. As Finishing Supervisor his job is described as essentially concerned with 'maintaining' production, and this term permeates his discussion about the Eight Step Plan teams. Peter has five turns in this extract and two of these turns are asides. The three substantive turns are concerned with 'maintenance'. As far as the Eight Step Problem Solving Plan group is concerned, his responsibility is one of 'maintaining the thing' (lines 22, 29) or 'maintain[ing] the group . . . just keep[ing] the work flowing'(line 29). Peter's Eight Step Team is not differentiated in any way from the other tasks he undertakes as Finishing Supervisor, it is all part of the 'amount of work that builds up it just goes to the back of the queue sort of thing' (lines 24–25). When James begins to construct the task of facilitating as a specific, new task, Peter interjects ironically, 'excuse me' (line 51), indicating that he rejects the 'workplace education' discourse in which James seeks to locate them as 'facilitators'. Peter seems to be firmly positioned within the local workplace discourse as a supervisor whose task is one of 'maintenance'—'keep the work flowing'.

(iv) Ben and Grace: being the team leader

The essential problem for Ben and Grace is that they are positioned by local workplace discourses in one set of social and power relations, and by global Quality discourses in another, contradictory set of social and power relations. They are seeking ways of 'being team leader', of resolving the dilemmas that have arisen as they have attempted simultaneously to occupy the roles of supervisor and team facilitator. While Sally and James rely largely on discourses around 'Action Learning' that seem to exist concurrently (and perhaps temporarily) in workplace education discourses and in Quality management discourses, to construct new identities for themselves, and Peter relies on familiar local workplace discourses involving supervisors and workers, Ben and Grace 'work' the local and the global discourses to construct 'alternate selves' that can function in the new/old work environment.

Ben explicitly names the problem as being one of social relationships and of knowledge. He suggests the solution lies in his facilitating a group where 'I'm away from the people a bit' (line 2), 'just have a background in what's going on' (line 5). When Peter interrupts to argue that it is a work-load problem (line 29), and James attempts to reword Ben's problem ('what you really want' line 31) as a problem about workers identifying as autonomous team members, Ben insists that it is a problem about relationships, and particularly a problem about power relationships, again calling on the metaphor of space and distance:

Ben	what I'm trying to get across
Peter	'cause
Ben	is I'm too close to those people because I already go outside of the group and then I'm their supervisor outside on the on the floor where maybe if I was facilitating another group where I'm not I'm not above them you know I'm not their supervisor or whatever (lines 31–37).

A metaphor that calls up the hierarchical structures of the local workplace discourse, Ben sees his task as 'just keeping them on track', a metaphor that Grace picks up and extends: I think we have to go down this path to try to direct people on to the path (line 11) suggesting that 'we' (the Team Leaders) must travel the path and that their task is to 'direct' people onto the path. Although Ben and Grace explicitly call on local workplace discourses for their power, they are ready to contemplate a situation in which the structures and relationships of 'Quality management' coexist with traditional workplace structures:

Ben	um I can go back to my job they can go back to theirs and they still um you know it's this their more their team than
Sally	yours
Ben	then everybody talking to the supervisor sort of thing you know

I've argued that Ben, Grace and James are 'working' discourses to form new working identities in this extract. One of the indications that they are actively engaged in this process is the elaborate hedging that occurs in their turns. In line 1 Ben introduces the topic with a series of hedges ('we thought, you know, maybe', 'or something'), and Grace picks this up in her expansion of the topic in line 9 ('Well, I think, kind of'), line 10 ('I mean'), line 12 ('we kind of') and line 14 ('us sort of just'). James also hedges as he searches for words to describe new workplace relationships line 27 ('sort of'), line 44 and 48 ('like', 'like'), line 53 ('somehow, you know, like, like'), line 71 ('you know like sort of'). They appear to be reluctant to be categorical, recognizing the transitory and contingent nature of working identities.

The Action Learning Team at AFM clearly is what it is intended to be, a potent site for the formation of new working identities through the textual practice of work. The process of identity formation is not, however, a straight forward process of 'colonization', and it does not produce a homogenized 'autonomous worker'. In this meeting at least, identity formation is dynamic, and the 'take up' of new identities, and the relinquishing of old identities, unpredictable and contingent. Each of the participants in the meeting 'works' the discourses washing through the site to form working identities which accommodate (to a greater or lesser extent) Quality management discourses but which are also influenced by the range of persistent workplace discourses that permeate AFM.

The textual production of working identities

Economic globalization is generally taken to be a system of global economic production, a system which relies on the standardization of textual practice to control work practice and homogenize working identities. Global companies dominate this system. In order to achieve comprehensive control at remote sites, global companies seek to colonize (and homogenize) the personal identities of workers and so achieve comprehensive peer and self-surveillance, forms of surveillance which can effectively replace more obtrusive traditional forms of control. In the rhetoric of global management discourses, workplace education occupies a strategic place in the formation of these new working identities. Learning has already become 'the new work practice' and Workplace Educators are charged with the responsibility of inducting workers into the homogenous cultures and practices of the generic learner, the 'autonomous professional'. It is tempting to conclude that workplace language education has been effectively co-opted to the service of global companies and global economic systems.

But this is not precisely what is happening as Ben and Grace and James and Peter and Sally sort out new working relationships and new working identities. For a start, they ask to have the meeting at Harbor Mill, on the Warping Shed Floor. Perhaps this is because the persistent demands of existing work practices can more clearly be articulated in the context in which they are carried out; the pressures they are experiencing can be better understood in the shadow of the warping machines. In this context they are better able to *improvise*, to create new ways to talk about work and to use the meeting for their own purposes.

Economic globalization is a *social* accomplishment achieved, moment by moment, in textual practice. Workplace textual practices are the microprocesses of globalization, and workplace texts realize the working identities and social relationships (both local and remote) of specific workers (like Ben and Grace and Peter and James) at local workplaces (like AFM) as they engage with global systems (like Autoco). Economic globalization cannot be understood independently from the people and the practices that constitute it.

What is significant about the texts analyzed here is not their homogeneity but their hybridity. Implicit in the *Autoco Quality Manual* is a bid for a new kind of worker, a worker who engages in autonomous, but absolutely predictable, problem-solving processes. Workers at AFM are not blank slates, they are part of personal and institutional histories, and they bring these resources to the processes of identity formation. While they must formally meet the demands outlined in the *Autoco Quality Manual* they do not have to simply 'take up' the working identity offered to them, indeed it would be almost impossible for them to do so. Ben, Grace, Peter and James, and Sally, co-opt the meeting to produce not one but several working identities, identities which variously incorporate the values and ideological positions which underpin Quality management discourses but which are also shaped by the persistent discourses of their workplace and their personal lives.

The place of workplace education in this whole process is critical. In this case at least, workplace language education is a 'site of struggle' around the formation of working identities at AFM, and this struggle is played out in the emerging textual practices of work. In the next chapter I look at this process from another point of view, from the perspective of workplace educators who find themselves in the paradoxical position of gaining power and losing control in contemporary workspaces.

CHAPTER SEVEN

Workplace educators working knowledge in the Knowledge Economy

> General notions of 'lifelong learning' have become increasingly colonized by discourses of 'human capital', 'competence' and 'total quality', producing a strange ideological brew merging human development, profit and productivity.
>
> (FENWICK)

INTRODUCTION

The workplace educator has been a more-or-less silent presence in this book so far, and yet the work of workplace educators like Sally, Margaret and others is often the catalyst for the reconstitution of knowledge across global networks and within local sites. In this chapter I want to talk about who workplace educators are, what they are doing when they embark on workplace or workforce education programs, and what they think about it.

As I write about workplace educators I am conscious that their established working knowledges, work practices, and working identities, are under at least as much pressure as anyone else's in the Knowledge Economy. They are 'working' knowledge in local workplaces and global workspaces, textualizing it and therefore reconstituting it, often without being fully aware of the contexts in which they are operating or the implications of what they are doing. Workplace educators teach people to read and write and talk in new ways, ways demanded by organizations and global regulatory frameworks, rarely by the students themselves. When they do so, they are challenging established status and power relations and sometimes the whole social structure of local working environments.

Sometimes workplace education seems unambiguously to be aimed at detaching knowledge from the people who know it. When workplace educators teach people to read and follow the Manual from Head Office under all circumstances, they seem, in a painfully paradoxical way, to be pivotal to a process of individual 'deskilling' that Braverman identified thirty years ago. Embodied and embedded knowledge is becoming almost invisible and inaccessible, and, increasingly, legitimate knowledge is seen as being located in, and generated by, the text. Sometimes it seems that the only skill a person really needs is the ability to read and write in specific ways, ways that make their hard-won knowledge available to anyone and their individual work practices reliant on what 'the manual' directs. At other times the work of workplace educators seems much more securely focused on the more complex and significant task of 'making knowledge common', on developing an orientation to textual practice that allows the respectful collaborative production of new working knowledge.

Workplace educators are critical to the development of the Knowledge Economy, and in many ways they are more important than they have ever been, and yet many feel they are negotiating a minefield. In this chapter I look at the experiences of seven workplace educators as they try to make sense of the work they are doing. My argument is that a (perhaps even *the*) critical function of workplace educators is to reconstitute knowledge at local sites in ways that make a global workspace possible; workplace educators are fundamentally discourse technologists.

WORKPLACE EDUCATORS AS DISCOURSE TECHNOLOGISTS

Fairclough (1996) argues that the global economy demands discourse technologists, people whose job it is to intervene in the textual practices of local sites in order to shift discursive practice to standardized global discourses. He understands discourse technologists to have access to special kinds of knowledge, reified knowledge that comes from outside the organization. Reified knowledge is decontextualized and presented as non negotiable. In the case of workplace educators who operate as discourse technologists, they may not know what the knowledge they present will mean when it is animated in the local contexts in which they impart it.

Discourse technologists do not simply teach specific discursive practices, they research the discursive practices of institutions, design discursive practices in line with institutional aims and strategies, and train people in their use. From this perspective, when Sally and Margaret try to teach Bill to problem solve with the Eight Step Plan, or even to contribute to a meeting where formal agendas are issued and minutes are recorded and archived, they are acting as discourse technologists. Their work is unambiguously aligned with global economic interests and objectives, although, as Sally explains below, this may not be immediately obvious. This doesn't mean that local knowledge and local discursive practices are no longer important in local workplaces, they are. It does mean, however, that local knowledge and local discursive practice is mediated through global discursive practices and that this has implications for how texts and practices are interpreted and valued at local sites.

In general terms, the function of the discourse technologist is to shift the control of local discursive practice from the local to the global. First, they shift the *policing* of discursive practice from the local to the transnational level and, second, they mediate local and global discourses at local sites in order to shift *legitimacy* to remote authorities. Workplace educators are prototypical discourse technologists, shifting legitimacy from the local to the global, realigning working relationships and reconstituting working knowledge and work practice in the process. This is not always a position they fully understand, nor one they seek. They often find it both surprising and profoundly disturbing, and seek ways to ameliorate what they perceive to be the negative impacts of their work while reinforcing what they hope are the positive impacts. While they do their work in local, physical workplaces, they are (often more or less unwitting) participants in globally distributed workspaces, recruited to the work of global regulatory systems, global corporations and globally distributed networks of production.

Workplace educators have a wide variety of educational backgrounds and professional biographies. They work at very different sites within institutions and global companies, as contractors or employees, and in many instances are at least partly government funded. No matter what their personal biography or professional location, workplace educators play a fundamental part in the microprocesses of globalization. Without them the textualization of working knowledge and work practice could not take place. For each of the participants in the project, their new prominence is taking some getting used to, raising concerns about who they are in the workplace, and why they are there. Like the workers around them, their own work practice is being redefined, their own knowledge is being reconfigured and their own working identities are being reconstructed.

The workplace educators I talk about in this chapter were, with the exception of Sally who worked at AFM, participants in a study called 'Workplace Educators "Working" Knowledge in the Global Economy'. My aim in this study was to talk with people who had a formal educative role in workplaces to get a sense of what their work entailed, how they did it, and what they thought about it. I talked with people working in Technical and Further Education Institutes, people working in corporations as trainers or as consultants with a training role, and people working as trade union trainers. They offered programs in a range of industries—'old economy' industries including heavy manufacturing and 'new economy' industries like ICT and service industries. Ten years ago it is unlikely that a Trade Union Trainer and a Corporate Trainer would have found much in common in their work. Now they both teach people to manipulate texts.

The participants in the project came from a variety of backgrounds. Mike and Tom are Trade Union Trainers, employed by a large Australian communications union. They began work as apprentice electricians and moved into the union first as organizers and subsequently into a range of roles that led to trade union training. Jessica and Gail are Enterprise Based Teachers employed by a Technical and Further Education College and contracted to teach programs in workplaces. Jessica trained as a technical teacher specializing in secretarial studies but moved into post-school technical and further education many years ago. Gail trained as a secondary teacher in the United States, worked in secondary schools in Australia and moved to post-school technical and further education five years ago. Originally their work involved teaching and basic technical skills and they still do that, but they now also teach many workers who are technically expert (often in IT-related fields), are confident in English literacy, and have relatively high levels of formal education. Melissa is a Corporate Trainer. She qualified as a secondary school English teacher and moved into private enterprise as a trainer very early in her career. She works in the Human Resources department of an insurance company which has been acquired by a global financial services corporation. Alison is a private Learning and Development Consultant who previously worked in a senior role as Learning and Development Manager for a large corporation in the ICT field. Alison had no formal training past the minimum school leaving age but gained broad experience in a large corporation and became a manager before taking her first qualification, a Graduate Certificate in Training at a TAFE college. For each of them, their work involves training people in the textualization of knowledge and skill to facilitate knowledge transfer globally and to reformulate 'what counts' as knowledge in local workplaces

POLICING KNOWLEDGE AND SKILL

Workplace education can be the direct means of shifting the policing of discursive practice from the local workplace to an external authority. In many respects, shifting *policing* is a relatively simple task because it involves the explicit (and possibly brutal) exercise of authority. The experience Gail recounts mirrors the experiences of many workplace educators I talked with in the Workplace Educators Project. She describes a common situation. The company she is contracted to work in is in a process of change, and that change is the impetus for the specific training program. The company hasn't communicated this very well, if at all, to the employees who are undergoing the training; the training comes as 'a bolt out of the blue' to them.

Gail was managing a Workplace Assessor training program at a petrochemical plant that was undergoing major restructuring. The stated purpose of the training program was to train people to assess their coworkers' skills. Explicit skill assessment was a requirement of national and global regulatory processes. If the company was to integrate into the global organization, and compete for clients in a global market, it needed 'trained' workers. On the day I interviewed Gail she had heard that that the smoldering industrial trouble in the plant had erupted into a full-scale and ugly industrial dispute, and the training program she was involved in was at the heart of it:

> They've just locked them out. You might have heard on the news, they've just locked out 150 employees. So, I was asked first of all to write the material for them, and I wanted to do the first part of Workplace Assessor, with the backdrop that they were going to bring in national competency standards in the petrochemical industry, and the first step was to have assessors. These in-house assessors would be able to assess the skills and competencies of those already working there. Now, as you can imagine, a lot of the old workers didn't want to be reassessed-they're 'I'm fine, I've been doing it for thirty years', [but] there was this awful underlying thing which the men had heard about but really didn't know, and they saw this training as the first step. Which it was.

Gail and her colleagues were not aware that the company was planning to reduce its workforce as part of the restructuring, and she and her colleagues were shocked at the responses they received from the workers when they turned up to deliver the training:

> So, we had myself and three other trainers went in and did the actual training of these supervisors, they were shift team leaders. So, they were like Leading Hands. Well, it was the most hostile environment I've ever been in in my life. It was just unbelievably hostile. [It was as if they were saying] I know why you are here, you came to try and get to know the group before you start assessing.
>
> And we were an instrument of change, which we hadn't realized. We hadn't been fully briefed.

What the students in the class guessed or knew, but the educators contracted to do the training did not, was that the reduction of the workforce was to be based on workplace assessments of skill. Previously, skill assessment at this level had been an informal matter that had little to do with formal certification and qualification. In presenting the Workplace Assessment program Gail was introducing a new, formal, means of determining skill. For the first time in this organization, workplace assessment would be reliant on manuals, proformas and other written documents. Several older workers (some of them Leading Hands in Gail's program) knew that they would not pass the assessment if it was based on a written test. Their kind of knowledge was no longer the kind of knowledge that would count. While workers could not be made redundant on the basis of illiteracy, they could certainly be made redundant on the basis that they were not certified skilled.

> And the situation was also very harsh...this is the way industry operates, more so now than ever and particularly with an international company like [ ]. The hundred that were laid off, the first hundred that went-they arrived at work with their lunch and were ushered to a meeting place and...they were ushered by security men who were hired for the purpose-ushered to their lock up, picked up their overalls and, I mean, those men had nothing they could-they were operators, but they [management] were worried about industrial sabotage.

Gail was surprised and dismayed at the part she was inadvertently playing in developing the work practices that meant people (perhaps her students) would lose their jobs. She was used to seeing herself as a teacher, mentor and advocate, as contributing to the development of the workforce, with workforce assessment as a means of legitimating, acknowledging and building skills. She was not used to seeing herself in a policing role, aligned with an explicit corporate agenda. Gail and her colleagues were in this case able to reclaim some of their roles as student advocates and supporters, but in difficult circumstances:

> Someone would come and say 'You know, Mick and I got on really well and Mick worked next to me and Mick has gone and I know he's just moved out of his flat and I don't know where he is but if you could find out for me, if he makes con-

tact with you, could you let him know, and here's my address', and I had a book, a page that was getting names and addresses, and found myself getting names and addresses, and found myself in this position liaising between people and putting people in contact with each other.

The informal contact initiated by students who had not been laid off gave Gail and her colleagues the opportunity to complete the training for those students who had been laid off. When they expressed a wish to complete their certificates Gail and her colleagues finished their programs and the assessment off site. They devised ways of assessing skills that did not involve reading and writing. At least, they felt, they could provide the laid-off workers with the opportunity to gain certification for the skills they had, so that they may be employable in a new job market.

Workplace educators are not accustomed to seeing themselves in a policing role in workplaces. It challenges traditional relationships with their students: how does one negotiate a productive relationship with students when one may be responsible for making them, or their workmates, redundant? Indeed, how does one negotiate productive relationships with employers when relationships of trust seem to have been compromised?

LEGITIMATING KNOWLEDGE AND SKILL

Useful working knowledge is embedded in cultures, communities and systems; it is social, contingent and contested. To make 'what counts' as knowledge at any particular workplace seem normal and uncontroversial, to legitimize it, is the more difficult part of the discourse technologist's work. Shifting legitimacy involves investing authority in a remote institution (like national competency standards, or the *Autoco Quality Manual*, or ISO for instance), but it also involves obscuring the identity of the institution so that demands made by the institution are normalized, they seem transparently natural and right, just 'best practice'. Experts outside the organization not only police discourse practices (by, for instance, providing forms to report faults or protocols by which problems might be solved) but, through that process, shape 'what counts' as knowledge and who can say so.

To shift legitimacy is to disrupt the complex intersections of skill, knowledge and status that make up the social landscape of the local workplace. For workplace educators this can be a confusing and confronting situation. It may be hard to believe that an offer to increase a person's repertoire of textual practices (like teaching them to take the minutes at a meeting for instance) could be perceived as a threat, or that teaching people to assess skills (and therefore, presumably, validate them) could be viewed as part of a strategy to de-legitimate established forms of working knowledge. It is not always clear what is at stake when workplace educators work with global corporations and national and global regulatory bodies to make knowledge common. Established local knowledge and practices are neither necessarily bad nor necessarily outdated (although they may be both), but their 'localness' works against the consistency and uniformity that globally distributed workspaces seem to need. Generally, workplace educators see themselves as enhancing and increasing knowledge, rather than as reconstituting it, when they textualize it.

Sally, who works with the Action Learning Team at AFM, was acutely aware of the shift in her role over the period that she had been employed as a workplace educator:

> We come in here with our little jobs of you know doing a bit of literacy and language teaching. God, we are going to go out of here with quite a degree of experience in how to function in [management].

Sally identifies her own realignment away from the job of language and literacy teaching and towards a management function. She is explicit about the way she reframed the Action Learning Activities to meet the urgent demands of the *Autoco Quality Manual*. From the outset, and not surprisingly for a language and literacy teacher, Sally is eager to incorporate as much writing as possible into the Action Learning Team meetings, despite Baz's reservations:

> as a teacher and a trainer the method of delivery worried me right from the start. . . .it was like Baz wanted to save them the trouble of actually writing anything.

Sally aligns her work as a workplace educator directly with Michael's work as the Human Resource Manager and, by implication, with the agendas set by external agencies (like Autoco) that Michael is trying to satisfy. If the company doesn't meet the demands of these groups it will lose the custom of global automotive clients. Sally legitimized the Action Learning Team project in the eyes of some skeptical members of senior management by demonstrating how it satisfied the external demand for the introduction of, and documentation of, specific work practices. AFM is required to undertake a 'self evaluation' of its compliance with the *Autoco Quality Manual* prior to being evaluated by Autoco. The results were disastrous; if AFM could not do better they would lose Autoco's custom:

> Tony had the Autoco Quality Manual with him and Michael was all hot to trot out with this stuff because they'd done their self evaluation based as customers and suppliers . . . they came out at 28/50 which is not good.

So I kind of told him what we're doing and what we're up to and so on. But some of the questions in this questionnaire were to do with continuous improvement teams, and you get one point if you are [using them]. You get another point if those meetings kept regular minutes. Well, they're all doing that, so that's another point. Another point if there are actions taken as a result of those meetings. Well that was very much on this proforma that the Eight Step Teams are working through. Both the original Eight Step Plan and the guide that I gave them to fill out each meeting is very much to do with who is doing what. By the end of the meeting you know who's doing what and at the beginning of the next meeting everybody reports back on that lot.

The emphasis shifts from the learning that is going on to the points that can be gained by adopting certain textual practices. The recording of activities (the minutes, the problem-solving guide, the written Eight Step Plans), rather than the activities themselves, becomes the primary aim. Sally is acting as a discourse technologist; 'designing discursive practices in line with institutional aims and strategies and training people in their use'.

Jessica

For Jessica, too, workplace teaching has increasingly become a matter of policing the discursive practice of students and shifting legitimacy to a remote authority. Jessica is an Enterprise-based teacher at a Technical and Further Education College in Australia. Originally her work involved teaching adult literacy and basic education to immigrant workers and workers with little formal schooling. Increasingly, however, Jessica finds herself teaching workers who are by any common definition literate in English and who have relatively high levels of formal education. In fact she teaches many workers who have completed the final year of schooling and who sometimes have advanced trade papers and professional diplomas as well. Often the workers who join Jessica's class are there because they are high flyers, skilled workers the company wants to keep. Students in this category are working towards the in-house credential of a major multinational automotive company. The precise details of what is required to meet the requirements are determined by the Head Office of the company. Employees need to satisfy these requirements to be considered for promotion:

Jessica   It's called the [Company] quality—the [Company] Production Certificate.

Lesley   With the [Company] Production Certificate is there a requirement on the local company to have a certain number of their employees go through that?

Jessica   All of them have to go through it eventually. It's been introduced in the last twelve months and only a few have been handpicked to go through it. These people have been handpicked thinking they are—because they're are so bright-and the company handpicks the ones that have got the terrific ideas because they want to get the ideas from them, but they're a bit shocked at the number that can't do, fill the requirement and they're being very rigid about that requirement, which to my mind . . . if the guy is able to stand up and do a nice presentation . . . then surely that's enough, but no, if he can't do it the proper way he doesn't get the certificate.

Jessica's role is not to teach the technical part of the certificate, that is laid out in detailed and highly prescriptive technical manuals. Her role is to teach people to use a different kind of language, to generate and package their knowledge in centrally mandated ways:

Lesley   And the proper way is written in a specific format with specific headings?
Jessica It's quite sophisticated.
Lesley   What sorts of features are required in that to pass?
Jessica Well there's the graphs, there's all sorts, just anything that you need to demonstrate your solutions.
         . . .
Jessica And **why** we're looking at it, and what **is** the problem. What's the extent of the problem, so they need to have tally sheets to show how often this particular problem arises—or perhaps extract reports. They would access all of that. They have no problem in that, but it's just stringing it together to make some sort of sense.

Jessica is teaching people to be exhaustively explicit in describing context and entirely predictable and consistent in undertaking problem-solving techniques and in talking about them. She is teaching people, who have been selected for their capacity to solve problems, to dismantle their current, localized, problem-solving and communicative practices and replace them with externally mandated practices. People won't need to know each other to engage in collaborative work, they will simply engage in the mandated problem-solving practices.

Problems are a big issue in manufacturing, and more often than not they need to be solved by several people at different global locations. In Chapter Five I described Covisint's Virtual Workspace, the Business to Business exchange developed by automotive companies to support globally distributed problem-solving and knowledge production. The Problem Solver tool is addressing a large-scale problem:

> For every problem identified in an assembly plant, a manufacturer issues a problem case. In a typical assembly plant, industry experience shows that there can be anywhere from 15-50 problem cases issued daily. Extrapolating those numbers globally results in potentially over 3.2 million problem cases per year for the automotive industry. Currently, each case must be responded to in the manufacturer specified format. A supplier invests a significant amount of time and energy in the administrative tasks of answering problem cases. This time could be better spent in preventive quality planning to avoid the problems in the future.
>
> Covisint's Problem Solver tool provides customers and suppliers with a web-based means to communicate problems and prompt proper permanent corrective action plans from one central, individually secure, hosted location. The tool provides an industry standard methodology to respond to problem cases (covisint.com/solutions/qlty/pshtml 14/01/02).

The aim of Covisint's Problem Solver tool is to standardize problem solving practice, but, especially in a globally dispersed, computer-mediated environment, that seems to require standardized language practice. Jessica's role in teaching the company Production Certificate is to standardize textual practice in the workplace at a very high level—not just formulaic textual practice, like memos and minutes, but knowledge-creating textual practice which is by its nature innovative. The standardization of textual practice is seen as a way of creating a global community of practice where it is textual and technical practice that links the community rather than values, experiences, or shared stories.

Mike and Tom

Like Jessica, Mike finds himself in an unfamiliar, and not entirely welcome, position when he undertakes workforce training. Whereas once the union played an active part in negotiating what counted as really useful knowledge, and what it was worth in terms of wages and conditions, he is now required to teach externally mandated curriculum and to assess according to externally mandated assessment tasks, and there is no prospect of negotiation.

Mike and Tom are Trade Union Trainers for a large Australian communications union. The union covers workers in the telecommunications industry and their membership was, until relatively recently, concentrated in a single large telecommunications company. Initially a wholly government-owned utility, the company is now partly privatized and partly government owned. The company was the kind of company that young people joined on leaving school, undertaking all their training internally and expecting to spend their working lives in the company (there were, after all, no competitors when they joined). As the company privatized it shed staff so that, in a decade or so the company had gone from employing about 90,000 people to employing about 35,000 people. This has important repercussions for the union which found that most of its members were no longer employed by the company; indeed, many could not find work in the industry in Australia. The membership base of the union shrank dramatically. Like unions around the world (Payne 2001), this union found that, to retain and attract members, it had to offer services attractive to unemployed and displaced workers who are no longer bound together by traditional forms of solidarity, occupational membership or group allegiance and who are likely to be engaged in short-term and contract work with a range of employers.

Trade Union Training is an established function of trade unions, and it seems a natural progression to expand that function to include formal skill development and accreditation. Within a similar context in the UK, however, Payne argues that this is in fact a fundamental change in the orientation of unions to learning and education:

> an analysis of current learning initiatives by the TUC and individual trade unions illustrates the beginnings of a process designed to reformulate what is to be understood as trade union education and secondly, illustrates a significant increase in the importance and role of learning initiatives for trade unions and their members ... 'What is required are transportable skills that are analytical and knowledge-based (Payne and Thomson 1998: 27; in Payne 2001: 388).

To survive, it seems, unions must train their members to be independent entrepreneurs.

Mike and Tom are responsible for the training of workers who have left the telecommunications company with an exceptional training record but no formal, externally recognized, qualifications. If their skills are to be 'transportable' then they need to be made visible by accreditation. The union runs an employment agency which places members in domestic and international positions. If members are to be placed, then they need externally verified accreditation. The union, therefore, takes responsibility for policing the accreditation, teaching the externally determined program and then assessing members' competence. Tom runs the training section of the union and strongly supports the union's work in accrediting members:

Tom   And now they have a piece of paper that they can take from company A to company B, to company C and say "this is me, I am qualified and accredited and I can work anywhere in Australia" and for that matter overseas as well. Because there has certainly been an issue we dealt with for quite a while, we were looking at putting some people into Saudi Arabia to do some work and the telecommunications people that we have here who would be by designation in the Telco either senior technical officers or principal technical officers are not employable in Saudi Arabia because they don't have the title "Engineer".

Internal role nomenclature is not sufficient, members need a title that international employers understand, and that title is 'Engineer'. From Tom's point of view, it is the title, and not the knowledge and skill itself, that is the problem for unemployed union members:

> But they have all the skills sets, all the training, but one word stops them from getting a job. So they could go over to Saudi Arabia and earn US$100,000 and something thousand a year in a job which they would do with consummate ease, because the Telco still is probably the only communications company in the world that can build a telecommunications system from the ground up. There is no other company that has got the capacity to do that. So these guys have that skill but because they're called Principal Telecommunications Technical Officers or whatever and not Telecommunications Engineers they're unemployable which is just crazy stuff. So what this whole process of accreditation is about is to say "righto, we want to have a very measurable and marketable skill set, skill recognition, so that they can take a piece of paper and go to any employer anywhere in the world and say this is what I'm qualified to do, this is where I stand from that point of view".

Here, formal accreditation seems to be the issue as far as Tom is concerned. Later comments, however, suggest that there is more to it than that. While Tom sees the process of accreditation as a simple matter of going through a process that will recognize and legitimate existing knowledge and skill, the process has not been as straightforward as he had hoped. Some union members who have participated in the union training and have undergone the assessment process have failed the assessment repeatedly. A major, and unexpected, part of the trade union trainer's work is to take tutorials that coach participants through the externally mandated test. Tom explains what is involved in the training and what, from his point of view, the real crux of the matter is:

Tom   a large part of this tutorial that I just mentioned earlier . . . one of the things that we've had to spend a lot of time with some of these guys on . . . is to teach them how to actually utilize a resource. I mean they have manuals and they need to be able to, not just look at the manual, but they need to be able to actually index it, and put little "post-it notes" on it and say OK that's a particularly critical part. And so they can refer back, because the exam they do allows them to refer back to the manual. So, it's very much a process of saying, OK this manual is a critical element, and we will teach you how to use this manual as a proper resource rather than something you just have sitting beside you or you keep in a bag which you hardly ever open from one week to the next.

Here Tom presents the training as a matter of adding new skills to an existing repertoire. He views the training as being fundamentally concerned with 'learning to use resources'. When participants in the Trade Union Training tutorial learn to read the manual and annotate it with Post-it Notes, and use it for the exam, they are, of course, learning new ways of reading and using written texts, and new ways of writing and presenting their knowledge. The requirement that they read and write in these ways does not come from their union, nor does it comes from the industry. Neither the union nor the participants are in control of the curriculum that is taught or the ways that it is assessed.

Payne, while acknowledging that 'there is an important issue of curriculum control here which should not be ignored (: 388)', argues that the control issue is essentially about how the different groups (the employers, unions, government-funding agencies and policy-makers) negotiate a compromise of their individual interests. This may, however, be a relatively benign reading of the issue of curriculum control. The participants in Mike's class are being asked to relinquish local knowledge that relies on personal networks and allegiances, where knowledge is embedded in practices and systems and cultures, and to take up a new set of practices where knowledge is located in the text (in this case the manual). In training these workers to 'use the manual' instead of their colleagues, Mike is engaged in building a new professional network, one in which the common practice described in the manual provides the connective tissue and personal relationships are virtually irrelevant. To put it another way, an important effect of the credentialing is to disengage the workers from their established professional community and to assert the primacy of externally controlled knowledge. As Tom, says, the question becomes:

How do you utilize this resource, turn this thing [manual] into a constant companion and friend . . . so that you can stroll through the exam?

In theory at least, the manual takes the place of the mate.

Unlike Tom, Mike is very uneasy about the whole process of teaching and accreditation. He is experiencing painful dilemmas in shifting from teaching traditional trade union training, in which he determines the curriculum, to delivering vocational education modules developed by employers, governments and educational funding authorities. Externally mandated training programs do not incorporate the kinds of critical thinking Mike identifies as integral to union work, and he finds it difficult to incorporate them. He talks of feeling compromised:

> Mike   I want to control the process. . . . you really need to try and empower the workers to take control of their learning processes and to give them that critical facility to look at things critically, to argue. A lot of that gets compromised in the whole process of trying to deliver the training, or deliver the skills at the same time. There is a role that could be pursued, but, by and large, it ends up being subsumed by our own particular requirements in this whole process.
>
> It was much easier when we were doing trade union training because then you had the luxury of just talking about the things that people needed to combat, their supervisors and this machine that was trying to crush them. But when you're tied up with a system of vocational education where there are training packages and there are certain things laid out in this training package it makes it much more difficult then to find an area to deaden that, to sort of try and develop some of these ideas and what have you. It's nice in theory but sometimes hard in practice.

Mike is aware that he is not simply teaching union members new skills, he is also teaching them that 'what counts' as knowledge has changed. In his early forties, he is dismantling two local communities, each of which he has belonged to since he was sixteen. One is based on well-established, tacit, embedded workplace knowledge and culture and one is based in the explicitly collective relationships of union solidarity. In their place he is helping to construct a workforce where the importance of any kind of personal relationship of trust and connection is minimized and the importance of centrally mandated, textually-mediated, practice is promoted as the only knowledge that counts. The possibility of building relationships, of solidarity or collective action, in this new alliance is small.

Alison

Alison is a Learning and Development Consultant employed by a multinational consulting firm. She is contracted to undertake specific training roles in multinational corporations. Often, the organizations she is invited to work in are restructuring, and the restructuring is the impetus for the training. Staff are being made redundant or are competing for positions within the company that they have held for several years. Previously, their knowledge and skills had been accepted and acknowledged; now they are being asked to articulate and justify their knowledge and skills and, often, reinvent themselves as a new kind of person, the kind of person the restructured company now requires. And they do this by engaging in a new kind of writing.

The focus of this reinvention is the 'resume', a document that is intended to capture who a person is and what they can do and communicate it to prospective employers. In a large and complex organization, a resume is a convenient way of comparing employees and fitting them into new job structures (or 'letting them go'). Previously, employees may have assumed (possibly incorrectly) that their knowledge and skills were obvious, at least within their own workplace. Now they cannot take this for granted but must learn the skill of creating a written working identity that matches the new company profile. This is the training task that Alison is involved in:

> So it was very difficult to go in and write a resume, just say, well, this is a resume-writing course, because there are so many issues that weren't dealt with. . . . They didn't know what their skills were. They didn't know how to articulate those into, they didn't even now how to articulate them verbally, never mind to actually write them in a CV. They didn't know where they wanted to be in 5 years time.

While teaching people to write a resume is a writing task:

> It is never purely a writing task and that, this is the challenge of my job. Because it looks like a writing task if you are a line manager and you think you've been very benevolent to give people something, and you are going to pay quite a bit of money to do that 'cause I don't come cheap and so they will get me in and there is no way that it is ever simply a writing task.

For Alison, part of teaching people to write a resume is teaching people to inhabit the identity they create:

It's a written exercise. We might just give them a sheet of paper with some action verbs, which is kind of a bit odd because people always say that verbs imply action. But people very often think in terms of fairly low level verbs like 'assist' or 'co-ordinate' or 'review' or ' I liaise a lot as well'.

So what we are starting to say to them is 'Here is a sheet of paper, and these have got some action verbs on them here, look at something that packs a punch, something that has got some power behind it. And describe yourself in those terms of what you do on a day-to-day basis. When you are doing what you do, right from the time you come in and you turn on your PC, if you do that, what is it that you are doing and writing that down and then going back over it and refining it and looking what verb might be more powerful to explain what you do from that point of view. And the personal skills that we look at are, are you self motivated or articulate etc, etc, etc. So, what we are hoping to do, what I tend to do a lot of this for, is so people start to see on paper, okay I am a bit impressive . . . I haven't actually wasted my 20 years, I've got some skills here that I can actually so something with.

And so, it is only then that you can, that any of us can really put a resume together and that is going to sell, going to personally sell us in the market place.

Melissa

Alison's work involves legitimating local workplace skills. She helps people see the practices and processes of their working lives in terms of knowledge and skill that can be reframed in ways that make it visible. She is trying to find ways to identify and reframe people's local knowledge in ways that count on a highly competitive, always shifting, global stage. For Alison the struggle is to help people see what they do as involving knowledge and skills that can be utilized in other environments, it is to legitimate local working knowledge. For Melissa, the task as she sees it is to de-legitimize local working knowledge in a context in which 'what counts' as knowledge in the local workplace would be held up for ridicule, and leave the global company open to litigation, if it went unchallenged. Melissa, like Alison, is a Corporate Trainer. She works in the Human Resource department of an insurance company which had recently been acquired by a global financial services corporation. Her work involves developing and delivering highly specific training to different groups within the organization.

It is easy to romanticize local workplace networks and communities, suggesting that shared cultures and relationships of trust, must inevitably produce valuable working knowledge. Existing, local, working communities are not, however, reliably benign. In Chapter Three I outlined some of the ways in which 'what counts' as knowledge is socially produced, and reflects and reinforces prevailing social attitudes and values. Melissa became aware of the implications of her role as workplace educator when she tried to introduce the textual practices and standards of a global finance company and came up against an exceptionally resilient local workplace culture.

A major function of claims assessors in the insurance company is to take statements from claimants. These statements, and subsequent records of telephone and face-to-face interactions, become part of the claimant's file. They are stored on computer and are available to any employee of the global company with access to the file, and, in most cases, to the claimants themselves, should they be inclined to request them. When she arrived at the company Melissa had initially been startled by the kinds of comments people wrote in customer files:

> a customer [is] on the phone putting in a claim and they weren't happy with the decision and he wrote on, he typed in "not a happy little vegemite". Now apart from that, its harmless you'd think, except the person was black. When they read their file—because they had access to their information—so he took [the company] to court.

New management was, perhaps unsurprisingly under the circumstances, happy for her to run a program that addressed the question of what could be said about a client, and what could be written down, in the new corporate environment. She reported that she was unprepared for the fury this unleashed when she presented her training program:

> and so I started to develop this . . . actual training program and I made it—they wanted it in short and sharp modules. So I made it 2 hours and all the organization was going to be trained up in it, but I targeted supervisors and managers as the trainers of this. So I've been doing a train the trainer.
>
> . . .
>
> I thought it was just common sense—I got examples from our own system to show how not to write. I used real examples, I condensed them, I changed names that sort of thing. So, if people said to me "oh but people wouldn't write that" I could say "well, yes they did". And it was interesting, because people were writing things like 'the customer is a liar'—'the customer lied about this'—and um 'the customer was not emotionally stable'.

Melissa was working with 'the claims end', supervisors and managers of operators who took telephone calls from clients whose claims had been denied either because they are believed to be fraudulent or because the policy they have does not cover them for the event.

> We do here [in the company] what I suppose police do in other organizations—we target profiles because its our way of risk assessment and so what happens is-I think through lack of understanding of research and statistics, people form what I call silly conclusions. I'll be blunt. I mean one example is 'you don't trust Arabs'. Anyone with an Arab sounding name—why? Because they've had a couple of experiences where there's been like a scam, you know.

Melissa developed a training program that focused on the written word and the written statements people had actually made:

> so people like put the mute on and go 'I've got an Arab on the phone' and it determines then how you actually treat people. So my training program was about generalizations, assumptions, what objectivity is, the impact words can have—that sort of stuff-and actually taking their statements. I actually was quite—I wasn't going to pull any punches—I took their statements as examples and activities and got them to work through—'is that true?'—'is that true for all the Arabs?'.

> I got 12 people together you know, and people that are considered high potential, thinking that I'll be able to be critical about this, but constructive, and you know give me some ideas about how to do it. Well I didn't even get a chance to get into the training program. It was meant to run for 2 hours. It went for two and a half to three hours debating where what I thought were reasonable adults telling me that it was OK to write 'the customer lied' and if the customer lied and the facts tell you that it's OK. And I said "but how do you know the facts tell you that?"—"that's a conclusion you're drawing" and then it got into the debate—we're a Plain Language company—but that's for our policies. They said "we're Plain Language—why do you want to flower things up and you know and avoid the point of saying the customer lied or is a liar?" And I said Plain Language isn't about insulting people. That was my point.

Melissa's story is instructive partly because it demonstrates the resilience of many local workplace cultures and partly because it shows how intimate and compelling are the connections between the values and orientations of local workplace cultures and the knowledge they produce and maintain. For Melissa the training program, and the fallout from it, had significant ramifications. She had to seek the support of the most senior people in the company:

> I hadn't anticipated for something which I considered common sense and just—small—got so huge where I had to involve the CEO and corporate lawyers, you know.

Melissa reported that, while some were convinced, many managers and supervisors remained hostile to training about what could be written on files, who had the right to access files, and what could be said and written about different social groups. They and the operators they supervised clung tenaciously to their beliefs (understood as facts) that beliefs about the propensity of certain social groups to commit fraud constituted knowledge, and that this knowledge was part of what made them effective claim assessors. Nonetheless, their practice had to change. Their files could no longer be considered private, belonging only to them, nor would they be read only by people who shared the views and values. What counts as legitimate knowledge, knowledge that could be written down and used, has been reconstituted, even if what counts as knowledge in the local workplace is harder to shift.

CONCLUSION

When workplace educators textualize knowledge they change it, and inevitably they get caught up in the political and social negotiations of local workplaces. When they textualize it according to practices mandated by authorities external to the local workplace—national regulatory frameworks, global companies, global standards—then they both police and legitimate conflicting discursive practices and the conflicting constructions of knowledge embedded in them. This process must change the social landscape of local workplaces and the working identities available. It will inevitably elicit hostile responses that challenge the ways that workplace educators think about themselves, their work and their knowledge.

 Afterword

IF WORKPLACE LITERACY EDUCATION IS THE ANSWER, WHAT IS THE QUESTION?

There were six of us around the table. Mike managed a technical textiles factory, Aaron taught ESL in workplaces, Susan ran the enterprise-based education programs for a local TAFE college, Cassandra was research project director for the union, Jack ran the community-based literacy programs for a local neighbourhood house, and there was me. Thirty people were being retrenched from a workplace, and there was a small amount of research funding available to develop an innovative, 'best practice' workplace, education program. According to the government funding sources, workplace education was the answer, but what was the question?

The retrenched workers were in their early to late forties. They had been working together for over fifteen years in loose, overlapping groups of six or seven. Their employer was adamant. They were good workers, knowledgeable workers. He knew them, liked them and trusted them. He was loathe to let them go. But the company relied for its survival on one major client, a global automotive company. They were developing new manufacturing techniques; their requirements had changed, and there would no longer be work in that section of the factory once the current contract was completed. The company had no choice but to respond immediately to decisions taken elsewhere. It had another worksite, on the other side of the city, but the retrenched workers were reluctant to travel four hours a day (especially either side of a 12 hour shift) to an unfamiliar place, for unfamiliar work, with unfamiliar people. Especially if the job was likely to vaporize, as this one had.

There is more to be said about this group of workers. Some of them are Vietnamese migrants. They have worked in one factory since arriving in Australia; they have learned their working knowledge and their work practices and have made their friends there. It has always been a good place to work, harmonious and flexible. They read and write in English, and in Vietnamese, and sometimes in French or Russian as well. They don't speak much in English, at least, not on the job. Some of the others are Greek or Italian immigrants. They, too, have spent most of their working lives in this factory. Their spoken English is fluent (and colourful). Most of them rarely read and write in English on the job and they speak Greek or Italian but don't read or write it much. Others are Anglo-Australian, they speak English (and some Vietnamese, Italian and Greek phrases they have learned on the job) and read and write it. Together, these workers have developed work practices that suit them, together they can solve problems, innovate, improvise, get the job done quickly and communicate effectively with each other and with management. Together, they are a skilled, knowledgeable and effective workforce, but they aren't going to be together for much longer.

There is more, too, to be said about this work place. It has been, by and large, a good place to work. It retains good workers because they like it there. Good work is recognized, innovative problem solving is encouraged and praised and management lets people get on with their work without too much interference. The people work collaboratively and their distinctive strengths are incorporated into the work rhythm of the team, and their weaknesses are accommodated. Everyone is encouraged to learn and develop new skills. Generally speaking, it hasn't mattered too much if an individual worker didn't read and write, or even speak much, in English, as long as the team got the work done on time and on budget. Generally speaking, it seemed good enough that management knew what people could do in terms of established and new skills, and paid them for it. Until now, formal skill assessment and certification seemed like an unnecessary burden on workers whose English literacy skills did not appear to extend to writing it all down, and on management, who did not need more paperwork to manage. Now, of course, none of us are confident about this.

We had come to the table because we had some workplace literacy education to dispense. We had agreed that workplace literacy now inevitably incorporated IT in this context—it was getting too hard to distinguish them on the floor in any case. The government was funding the development of a program, and their employer (he was a good bloke and overwhelmed that this had happened, and he could do nothing to stop it) was providing the hours in work time. So, what was the question that workplace literacy education might begin to answer? What kinds of literacy skills does a person need to be employable on the global market? What kinds of literacy skills does a person need to be a knowledge worker? What kinds of literacy skills does a person need to be able to make their knowledge a commodity that is visible on the global market? How (now here was a question!) does a person recognize, isolate and describe their own skills when they have been developed and used together, 'in practice'? How does a person recognize, isolate and make explicit the skills they have in making and sustaining a community of practice? All in all, how do you prepare someone for work in the future? Forget the future. How do you prepare someone for now? For those of us sitting around the table the question was even more acute. With only twenty precious hours of guaranteed, fully funded, on-the-job training, how do you prepare anyone for anything?

Wrong questions, wrong assumptions, start again. These people did not have to learn symbolic-analytic knowledge from scratch; they already were, at least sometimes, knowledge workers. Let's start from there. Sometimes, perhaps, most of the time, their work was routine. Not always, but often enough, however, they made and used knowledge in the local setting of their working lives. As a group, located in the one physical space which they knew well and helped to create, they identified and solved problems, improvised, innovated and collaborated. If they hadn't been able to do this their jobs would have disappeared long before this. Robert Reich had already roughed out a curriculum for symbolic-analysts, for knowledge workers:

> The skills people need to develop have to do with problem solving and identification, developing critical facilities, understanding the value of experimentation, and the ability to collaborate. In other words, given that economies are changing so rapidly, the most valuable skills someone can acquire are the skills to learn rapidly and efficiently and to go into almost any situation and figure out what has to be learned (Morrison quoting Reich 1991).

A good, if brief, curriculum for the development of what is commonly understood to be the symbolic-analytic knowledge worker, the elite (the games designer, financier, journalist or doctor) but hardly adequate for everyday workers who make and use symbolic-analytic knowledge at work to keep global supply chains operating. It is one thing to be an elite knowledge worker who may be more or less assured of employment, as long as they can stay away from any work that threatens to become routine. It is quite another thing to be a routine 'knowledgeable worker' who is far more exposed to the vagaries of global markets and who, because of relatively low pay, is far more vulnerable to them. These workers might benefit most from a literacy program that helps them identify their skills in innovation, improvisation, problem solving and collaboration, helps them identify the written, digital and spoken texts they currently use to do this work, and helps them begin to imagine the texts they might need to do this work when they are not working together, in a physical workplace.

That's not all, of course, after all, we have twenty hours. These workers are vulnerable and the level of autonomy they may have in any particular network of interaction is unpredictable. If they were ever in a safe place, they certainly won't be in one from now on, and they need to be able recognize when texts are positioning them in ways they would rather resist or deflect. If texts are the contexts in which we do our work, then knowledge about how texts have worked up to now in a familiar environment, and how they might work in unfamiliar environments, and knowledge about how people work texts, might be the only absolutely essential knowledge a person can hope to have.

I am reluctant to speculate about how the Knowledge Economy might develop, or what might happen to specific jobs, sectors or industrial processes, or even to Bill and Grace and Sally and Mary and the others at AFM, and Tom and Alison and Melissa and the other workplace educators. I am reluctant even to speculate about will happen to the students who have been offered a quite unlooked-for twenty hours of workplace literacy education. For me it is quite challenging enough to try to find a way of paying attention to what is happening now, as people engage in the textual practices that make the Knowledge Economy happen.

For those of us sitting around the table, trying to figure out what kind of training is most urgent, I'd like to make this suggestion: If workplace literacy education is the answer then, maybe, the question we should address in our twenty hours is 'How, in global networks of economic interaction, do we make knowledge, and make it common, and what is at stake when we do?'

References

Alvesson, Mats. (1993). "Organization as Rhetoric: knowledge-intensive firms and the struggle with ambiguity." *Journal of Management Studies* (90): 997-1016.

Arunachalam, S. (1999). "Information and knowledge in the age of electronic communication: a developing country perspective". *Journal of Information Science* 25(6): 465-479.

Barnes, J. and R. Kaplinsky (2000). "Globalization and the death of the local firm? the automobile components sector in South Africa". *Regional Studies* 34(9): 797-812.

Benner, C. (2003). "Computers in the wild: guilds and the next generation unionism in the information revolution". *International Review of Social History*, 48 (Supplement 11 Uncovering labour in information revolutions).

Berezin, P. (2003). "Did medieval craft guilds do more harm than good?" *Journal of European Economic History*, 32(1), 171-197.

Betcher, G. (2004). *Medieval Guilds*, Iowa State University. 2005.

Blackler, F. (1995). "Knowledge, knowledge work and organizations: an overview and interpretation". *Organization Studies* 16(6): 1021-1046.

Braverman, H. (1974). *Labour and Monopoly Capital.* New York and London, Monthly Review Press.

Brown, J.S. and P. Dugaid (2000). Organizational learning and communities of practice: towards a unified view of working, learning and innovation. *Knowledge and Communities.* E.L. Lesser, M.A. Fontaine and J.A. Slusher, Boston, Butterworth and Heinemann: 99-122.

Butters, J. and J. Bennett (2002). Covisint hits rough patch as business falling flat, Auto.com. 2004.

Castells, M. (1996). *The Rise of the Network Society.* Malden, Massachusetts, Blackwell.

Davenport, T. H. and L. Prusak (1998). *Working Knowledge. How organizations manage what they know.* Boston, Harvard Business School Press.

de Certeau (1984). *The Practice of Everyday Life*. Berkeley, University of California Press.

Duguid, P. (2000). "Balancing act: how to capture knowledge without killing it". *Harvard Business Review*, May/June: 73ff.

Engestrom, Y. (1999). *Expansive learning at work: towards an activity - theoretical reconceptualisation*. Seventh Annual International Conference on Post Compulsory Education and Training, Gold Coast, Centre for Learning and Work Research, Griffith University.

Epstein, S.R. (1998). "Craft guilds, apprenticeships and technological change in preindustrial Europe". *Journal of Economic History*, 58(3): 684-714.

Ezzamel, M. and H. Willmott (1998). "Accounting for team work: a critical study of group-based systems of organizational control". *Administrative Science Quarterly*, 43(2): 358-396.

Fairclough, Norman. (1992). *Discourse and Social Change*. Cambridge, Polity Press.

Farrell, L. (2004). "Workplace education and corporate control in global webs of production". *Journal of Education and Work*.

Farrell, L. (1999). "Reconstructing Sally: narratives and counter narratives around work, education and workplace restructure". *Literacy and Numeracy Studies. An international journal in the education and training of adults*, 9(1): 5-27.

Farrell, L. and B. Holkner (2004a). "Points of vulnerability and presence: knowing and learning in globally networked communities". *Discourse*, 25(2).

Farrell, L. and B. Holkner (2004b). *Working through ICTs in hybrid learning spaces*. Australian Association for Educational Research, University of Melbourne, Melbourne, Australia, AARE.

Farrell, L. and B. Holkner (2003). "Seizing the text: exploiting points of vulnerability in technologically mediated communications networks". *ERIC*.

Fenwick, T. (2002). "Canadian women negotiating working knowledge in enterprize: interpretive and critical readings of a national study". *Canadian Journal for the Study of Adult Education* 16(2).

Folinsbee, S. (2004). Paperwork as the lifeblood of quality. *Reading Work. Literacies in the New Workplace*. M.E. Bellfiore, T.A. Defoe, S. Folinsbee, J. Hunter and N.S. Jackson. Mahwah, New Jersey, Lawrence Erlbaum Associates.

Gee, J.P. (2000). "Communities of practice in the new capitalism" *Journal of Learning Sciences*, 9(4): 515-523.

Gee, J.P. (1997). "Beyond culture: communities of practice in the new capitalism. *Critical Forum* 5(1 and 2): 70-82.

Geisler, C. (2005). "Textual objects: accounting for the role of texts in the everyday life of organizations". *Written Communications*, 18(3): 296-325.

Graham, S. (1998). "The end of geography or the explosion of space?' *Progress in human geography*. 22(2): 165-185.

Granovetter, M. (1973). "The strength of weak ties". *American Journal of Sociology*, 78(6): 1360-80.

Hammer, M. (1996). *Beyond Reengineering: How the process-centred organization is changing our work and our lives*. New York, Harper Collins.

Hildreth, P. and C. Kimble (2004). Introduction. *Knowledge Networks: Innovation through communities of practice*. P. Hildreth and C. Kimble, Hershey, Pennsylvania, Idea Group Publishing.

Hildreth, P. and C. Kimble (2000). "Communities of practice in the distributed international environment", *Journal of Knowledge Management* 4(1): 27-38.

Hirschorn, L. (1984). *Beyond Mechanization*.Cambridge, Massachusetts, MIT Press.
Hull, Glynda (2000) Critical Literacy at Work. *Journal of Adolescent & Adult Literacy.* v43 n7 p648-52 Apr 2000.
Jackson, N. (2004). Introduction. *Reading Work Literacies in the New Workplaces.* M.E. Belifore, T.A. Defoe, S. Folinsbee, J. Hunter and N.S. Jackson. Mahwah, New Jersey, Lawrence Earlbaum Associates: 1-15.
Koch, C. (2002). Harold Kutner is Covisint's Last Chance, CIO.com.2004.
Lankshear, C., C. Bigum, et al (1997). Digital Rhetorics: Literacies and Technologies in Education—Current Practices and Future Directions. 3 vols. Project Report. Children's Literacy National Projects. Brisbane: QUT/DEETYA.
Lash, S. and J. Urry (1994). *Economies of Signs and Space.* London, Sage.
Lave, J. (1993). The practice of learning. *Understanding Practice: Perspectives on Activity and Context.* S. Chakli and J. Lave. Cambridge, Cambridge University Press: 3-32.
Lave, J. (1988a). *Cognition in Practice: Mind, mathematics and culture in everyday life.* Cambridge, Cambridge University Press.
Lave, J. (1988b). *Cognition in Practice: Mind, mathematics and culture in everyday life.* Cambridge, Cambridge University Press.
Lave, J. and E. Wenger (1991). *Situated Learning: Legitimate peripheral participation.* Cambridge, Cambridge University Press.
Lee, L. and M. Neff (2004). How information technologies can help build and sustain an organization's communities of practice: Spanning the Socio-Technical Divide. *Knowledge Network: Innovation Through Communities of Practice.* P. Hildreth and C. Kimble, Hershey, Pennsylvania, Idea Group Publishing.
Lesser, E., M. Fontaine, et al Eds (2000). *Knowledge and Communities.* Boston, Butterworth-Heinemann.
Lindstaedt, S. (1996). "Towards organisational learning: growing group memories in the workplace". *CHI 96.* M. Tauber: 53-54.
Lyotard, Jean-François (1984) *The Postmodern Condition.* Trans. Geoff Bennington and Brian Massumi. Minneapolis: University of Minnesota Press
Mann, G. (2002). "Race, skill and section in northern California". *Politics and Society* **30**(3): 465-496.
McLeod, Julie (1997). "Can we find out about girls and boys today or must we settle for just talking about ourselves?: dilemmas of a feminist, qualitative, longitudinal research project". *Australian Educational Researcher* **24**(3): 23-42.
Mitchell, William J. 1994. *City of Bits.* Cambridge: MIT Press.
Moslein, K. (2001). *The Location Problem in Electronic Business: Evidence from Exploratory Research.* 34th Hawaii International Conference on Systems Sciences, Hawaii, IEE.
Mukerji, C. (1998). The collective construction of scientific genius. *Cognition and Communication at Work.* Y. Engestrom and D. Middleton. Cambridge, Cambridge University Press: 257-278.
Nonaka, I. and H. Takeuchi (1995). *The Knowledge Creating Company.* New York, Oxford University Press.
Noon, M. and P. Blyton (1997). *The Realities of Work.* London, MacMillan Business.
O'Hara-Devereaux, M. and R. Johansen (1994). *Global Work. Bridging Distance, Culture and Time.* San Francisco, Jossey-Bass.
Orlikowski, W. (2002). "Knowing in practice: enacting a collective capability in distributed organizing". *Organization Science* **13**(3): 249-273.

Orr, J. (1990). Sharing knowledge celebrating identity: war stories and community memory in a service culture. *Collective Remembering: Memory in Society*. D.S. Middleton and D. Edwards. Beverly Hills, Sage.

Payne, J. (2001). "Lifelong learning: a national trade union strategy in a global economy". *International Journal of Lifelong Education* 20(5): 378-392.

Phillips, A. and B. Taylor (1986). Sex and skill. *Waged Work - A Reader*. F. Review. London, Virago: 54-66.

Pittard, M., R. Kowalski et al (2004). *Regulating the global workplace*. A paper presented to the Annual Conference of the Australian Labor Law Association, University of Sydney, April 2005.

Poynton, Cate. (1986). *Language and Gender: making the difference*. Waurn Ponds, Deakin University Press.

Porter, M.E. (1998). "Clusters and the new economics of competition". *Harvard Business Review* Nov-Dec: 77-90.

Prusak, L. and D. Cohen (2001). "How to invest in social capital". *Harvard Business Review*, June: 86ff.

Reich, R. (2003). Nice Work If You Can Get It. *Wall Street Journal*, December 26.

Scarbrough, H. (1999). "Knowledge as work: conflicts in the management of knowledge workers". *Technology Analysis and Strategic Management* 11(1): 5-12.

Senge, P. (1991). *The fifth discipline: the art and practice of the learning organization*. New York, University of Toronto PressDoubleday.

Sholtz, S. and M. Prinsloo (2001). "New workplaces, new literacies, new identities". *Journal of Adolescent and Adult Literacy* 44(8): 710-713.

Smith, D. (1999). *Writing the Social: Critique, theory and investigations*. Toronto, University of Toronto Press.

Snyder, W. (2000). "Communities of Practice: The Organisational Frontier". *Harvard Business Review*, January/February: 139.

Soja, E. (2000). *Postmetropolis Critical Studies of Cities and Regions*. Malden, Massachusetts, Blackwell.

Vanderschraaf, P. (Summer 2002). Common Knowledge. *Stanford Encyclopedia of Philosophy*. E.N. Zalta, http://plato.stanford.edu/archives/sum2002/entries/common-k.

Welch, D. (2000). E: Marketplace: Covisint, Business Week Online. 2004.

Wenger, E. (2005). Communities of practice: a brief introduction. 2006. http://www.ewenger.com/theory/communities_of_practice_.intro.htm

Wilson, Bruce 1997 The politics of valuing worker expertise: different approaches to organisational learning. *Melbourne Studies in Education*. 38 (1) 73-90

Zuboff, S. (1988). *In the Age of the Smart Machine*. New York, Basic Books.

Zuboff, S. and J. Maxmin (2002). *The Support Economy*. London, Penguin.

Index

-A-

action-dependence, 50
action learning teams, 75–77, 121–23, 126
Alvesson, Mats, 65
Arunachalam, S., 19
Australian Fabric Manufacturers, 28, 29, 31–34, 34–35, 35–36, 36–39, 41

-B-

Bakhtin, M. M., 20, 21
Barnes, J., 41
Bennett, J., 100
Berezin, P., 4
best practice, 9
Betcher, G., 4
Blackler, F., 46, 47, 48, 51, 64
Blyton, P., 49, 51, 54, 86
Braverman, H., 59, 134
Brown, J. S., 7, 19, 60, 67
Butters, J., 100

-C-

Castells, M., 10, 11, 15, 56, 57
city space, 28
Cockburn, Cynthia, 54
Cohen, D., 6
Collaboration Manager, 98, 101, 104, 105, 106
collective memory, 6
common knowledge, 4–5
    commonplace, 6–7
    common practice and, 5–6
    definition, 1, 7
communities of practice, 62, 63
context dependence, 50
continuous improvement regimes, 116
costing, 87, 88
Covisint Collaboration Manager tool, 96, 98
Craft Guilds, 4, 30
cyber texts, 104–106

-D-

data, 13
Davenport, T. H., 13, 45, 46, 47, 58
de Certeau, 67
deskilling, 134
digitized texts, 20
discourse technologists, 110, 134, 135
Duguid, P., 2, 3, 19, 60, 67

-E-

Edubase, 95
Ellison, L. J., 102
embodied knowledge, 47, 120
embrained knowledge, 47
emotion work, 54
encoded knowledge, 47
encultured knowledge, 47
Engstrom, Y., 6, 47
Epstein, S. R., 4
Ezzamel, M., 6

-F-

face-to-face communication, 17, 21
Fairclough, Norman, 87, 134
Farrell, L., 16, 94, 117, 119
Fenwick, T., 133
first space, 28
Folinsbee, S., 116, 117
Fontaine, M., 6, 47
Foucault, M., 31

-G-

Gee, J. P., 6, 20, 21, 62, 63
Geisler, C., 20
globalization, 26
global knowledge economies
    standardization, improvisation and, 9–11
    working in, 8–9

global workspaces, 26
Graham, S., 68, 91
Granovetter, M., 8
Graphic Artists Guild, 5

-H-

Hammer, M., 119, 120
Hildreth, P., 6, 47, 62
Hirschorn, L., 47
Hoffman, M. B., 102
Holkner, B., 16, 94, 117
HTML Writers' Guild, 5
Hull, Glynda, 70, 72

-I-

images of knowledge, 46
improvisation, 8, 10
incident outcome, 97
infinite replicability, 80
information, 13
Information Communication Technology (ICT), 12, 15, 16, 19, 46, 48, 52, 56, 57, 61, 63, 67, 68
iTexts, 20

-J-

Jackson, N., 80, 116
Johansen, R., 26
Just In Time, 119

-K-

Kaizen, 119
Kaplinsky, R., 41
Kimble, C., 6, 47, 62
knowing, 48
knowledge, 13, 48
    action learning teams and, 75–77
    free communication of, 1

in and with texts, 2
language and, 3
legitimating, 139–41
local, 4
new, 120
policing, 137–39
processes, 88
production of, 1, 69–72, 73–75, 77–88, 109
reified, 134
scientific, 64
social construction of, 51–55
social practices and, 2
textualization of, 59–63
textual practices and, 3
workplace educators and, 2
Knowledge Economy, 2, 3, 5, 7
knowledge in, 55–59
knowledge workers and, 11–16
working in, 8–9
workplace educators and, 133–34
Knowledge Management, 60
Knowledge Networks, 62
Knowledge Organizations, 58
knowledge production, 22
knowledge work
information and communications technologies and, 16–18
paperwork and, 116–18
knowledge workers, 2, 7, 11–16, 155
collaborative construction of, 63–65
politics of, 49–51
Kowalski, R., 13

-L-

Lankshear, C., 119
Lash, S., 9
Lave, J., 47, 62
Learning Teams, 119
Lee, L., 48, 63
Lesser, E., 6, 47
lifelong learning, 133
Lindstaedt, S., 6
Lyotard, Jean-François, 61
local knowledge, 4

-M-

Mann, G., 52, 5354
Marshall, A., 53
Maxmin, J., 15
Mitchell, William J., 92
Moslein, K., 25, 26
Mukerji, C., 64

-N-

Neff, M., 48, 63
new knowledges, 120
"Nigger-Killer," 52, 53, 54
node and nucleation, 38
nominalization, 87
Nonaka, I., 6
Noon, M., 49, 51, 54, 86

-O-

O'Hara-Devereaux, M., 26
on-line environments, 91
creating teams and, 93–96
organizational theory, 46
Orlikowski, W., 47
Orr, J., 6

-P-

PaceSetters, 94, 95, 117
Payne, J., 144, 146
personalism, 50
Phillips, A., 54
Pittard, M., 13
Porter, M. E., 25
Poynton, Cate, 86
Prinsloo, M., 42, 70, 71
Prusak, L., 6, 13, 45, 46, 58

-Q-

Quality Circles, 119
Quality Management, 119

-R-

Reengineering, 119
Reich, R., 11, 12, 155
reified knowledge, 134

-S-

Scarbrough, H., 45, 62, 63
school-based literacy, 70
scientific genius, 64
scientific knower, 64
scientific knowledge, 64
Senge, P., 6
sentience, 50
Sholtz, S., 42, 70, 71
Silicon Valley Web Guild, 5
skill, 51, 52, 53
Smith, D., 5, 22, 45, 47, 48
Snyder, W., 6
Soja, E., 9, 25, 27, 28, 31, 36
spatial forms, 31
Stanford Encyclopedia of Philosophy, 7
Sybase Inc., 102
symbolic analysts, 11, 12, 14
Systems Administration Guild, 5

-T-

Takeuchi, H., 6
Taylor, B., 54
Taylorism, 46, 59, 61
texts, 69–72
    collaborative problem solving and, 104–106
    micropolitics of, 87
    talk and, 88–90
    technologies and, 89
third space, 28, 43
transmissions, 22

-U-

Urry, J., 9

-V-

Vanderschraaf, P., 1, 7
Virtual Workspace Project, 99–101, 101–104, 106, 121, 143
    conflict and, 107–108, 108–111

-W-

Welch, D., 102
Wenger, E., 6, 62
Willmott, H., 6
Wilson, Bruce, 119
working knowledge
    conceptualizing, 46–48
workplace, 16, 17, 31, 58, 60
    educators, 111, 115, 131, 133–34, 134–37, 152
    knowledge in, 45
    literacy, 70, 154
    local, 91
    ruling relations and, 21–23
    technologies and, 18
    texts, technologies, identities and, 19–21
    virtual, 98–99
workspace, 16, 17, 18, 58, 60
    knowledge in, 45
    making knowledge common in, 55–59
    negotiating identity in, 122
    ruling relations and, 21–23
    texts, technologies, identities and, 19–21
    virtual, 98–99, 105
World's Best Practice, 119

-Z-

Zubboff, S., 15, 49, 50, 51, 61, 113